Bacterial Degradation of Organic and Inorganic Materials

Marziale Milani • Roberta Curia
Natalia Vladimirovna Shevlyagina
Francesco Tatti

Bacterial Degradation of Organic and Inorganic Materials

Staphylococcus aureus Meets the Nanoworld

Springer

Marziale Milani
Department of Materials Science
University of Milano-Bicocca
Milan, Italy

Natalia Vladimirovna Shevlyagina
N. F. Gamaleya Federal Research Center
for Epidemiology & Microbiology
Moscow, Russian Federation

Roberta Curia
Department of Biotechnology &
Biosciences
University of Milano-Bicocca
Milan, Italy

Francesco Tatti
Materials & Structural Analysis
Division (MSD)
Thermofisher Scientific
Eindhoven, The Netherlands

ISBN 978-3-031-26948-6 ISBN 978-3-031-26949-3 (eBook)
https://doi.org/10.1007/978-3-031-26949-3

This Springer imprint is published by the registered company Springer Nature Switzerland AG
The registered company address is: Gewerbestrasse 11, 6330 Cham, Switzerland

In memoriam

Preface

On December 8, 2015, Dr. Lyubov Vasilievna Didenko passed away. Since we were delivered the sad news, the desire to meet with all her colleagues to celebrate her memory as a scientist and a friend came upon us. The Memorial Day in her honor was held on April 6, 2016, at N. F. Gamaleya Federal Research Center for Epidemiology & Microbiology, Moscow, Russian Federation. The heartfelt memories of the people who shared with her work and everyday life triggered into us, among her closest collaborators (even though not in terms of distance!), the desire to spread her work, her personality and especially her precious contribution to the scientific community worldwide. This book is our present to Lyubov and her family.

Our collaboration with Dr. Didenko started almost by chance in front of an electron microscope, and from the beginning it turned out to be stimulating and promising. We worked together for several years, though too few, and presented our results in several papers and at scientific meetings.

The time spent in the laboratory with Dr. Didenko was a personal enrichment; Lyubov was not interested in just the mere scientific aspects of the problem under investigation, she also loved to think over and discuss the socio-cultural premises and consequences of the research activity, touching a variety of themes from science to humanities, actually building a bridge between the two cultures. Surely in doing so she was facilitated by her being a Medical Doctor and a woman of culture, who led a laboratory of advanced technology as the Laboratory of Microorganisms' Anatomy at N. F. Gamaleya Federal Research Center for Epidemiology & Microbiology, one of the world's leading research institutes with a solid and secular scientific reputation.

In our honest opinion the best way to honor Dr. Didenko is continuing the scientific activity following her inspiring principles.

In this book we will discuss the bacterial biodegradative activity on organic and inorganic materials and the consequences into the nanoworld; we will also talk about the key role of electron microscopy images and the relevance of their interpretation. A critical analysis of the scientific knowledge in the nanoworld, with an escape into the problem of perception related to the reading of images, will be presented without ignoring the hard questions that are almost always left to philosophers and intellectuals.

Advancing in the work started with Dr. Didenko, we cannot help but miss Lyubov not only as a great scientist but as a dear friend. Her memory will remain with us forever.

Marziale Milani
Francesco Tatti
Natalia Vladimirovna Shevlyagina
Roberta Curia

To those who did not have the pleasure to meet her we would like to present Lyubov's memory through the eyes of her husband, Dr. Alexander Georgievich Avtandilov,[1] who accepted to write the following cameo about his wife.

It has been a year since L.V. Didenko passed away, but the pain is not weakening, my thoughts remain with her, and I keep remembering these 32 years of bliss that we spent together. I cannot help but remember when we first met in 1982: I was a postgraduate; she was a resident at the Department of Infectious Diseases. She was a real beauty, with her dashing eyes changing color from gray to green in the summer sun. However, what amazed me most in her was her intelligence, her level of education that wonderfully combined with brilliant irony both to self and to the world around. Her gentleness and femininity melted my heart, and I gave up. She was a gift from God to me, an almost 30-year-old bachelor.

She was a wonderful wife and a selfless mother. She easily accepted the life experience and family traditions of our parents. She managed to succeed in everything, from delicious cooking and exemplary housekeeping to practicing science at her laboratory, which she took charge of when she was only 30 years old. The laboratory rose from the ashes and flourished under her lead, and became a landmark for many scientists from all universities of Moscow and the Russian Federation. The most bold and challenging theories that were developed there found their application in experiments and life. I.P. Pavlov's catchphrase "For the naturalist, everything is in the method" always played the major role in Lyuba's work. I believe that broad medical education, 5 years of medical practice, and constant self-development allowed her to take the leading position in science.

It is worth mentioning her huge methodological help to our son Georgy during his thesis research in biodestruction of polymer dental implants. In this study, it was for the first time showed how microscopy can display a correlation between different stages of bacterial colonization and structural changes in the surfaces of removable dentures. The study also defined the role of biofilms on the surfaces of polymer dental implants and developed the methods of their eradication.[2]

She did not stop thinking about her research even at home: she always discussed it with me and managed to come to the right conclusion. We always backed each other up: she was genuinely happy about my professional advancement, I was also

[1] Honored Doctor of the Russian Federation, Head of the Department of Therapy and Adolescent Medicine of the Russian Medical Academy of Postgraduate Education of the Ministry of Health of the Russian Federation.

[2] Didenko LV, Avtandilov GA, Shevlyagina NV, Smirnova TA, Lebedenko IY, Tatti F, Savoia C, Evans G, Milani M, "Biodestruction of polyurethane by *Staphylococcus aureus* (an investigation by SEM, TEM and FIB)", in: Méndez-Vilas A (Ed.), Current Microscopy Contributions to Advances in Science and Technology, Formatex Research Center, Spain; 2012:323–334.

very proud of her scientific achievements. This encouragement helped both of us to outlast the rough periods of life during the mid-2000s.

"Lyubov" means "Love" in Russian. She impersonated every aspect of love in her: love for our son, love for her husband, love for her friends and acquaintances. She was always able to find the right words for everyone. Lyuba bestowed her kindness, care, and sincerity on everybody.

I deeply appreciate the proposition of Professor M. Milani (University of Milano-Bicocca) and F. Tatti (Thermofisher Scientific—Materials & Structural Analysis Division (MSD), The Netherlands) to write a foreword to the book about my beloved wife—my closest friend and soul mate.

With tranquil sorrow and hope for reunion beyond the far horizon.

Alexander Georgievich Avtandilov
December 2016

Milan, Italy — Marziale Milani
Milan, Italy — Roberta Curia
Moscow, Russian Federation — Natalia Vladimirovna Shevlyagina
Eindhoven, The Netherlands — Francesco Tatti

Introduction

Lyubov Vasilievna Didenko was born on July 22, 1958, in Zagorsk (Moscow, Russian Federation) to Vasily and Catherine Didenko. Her family consisted of her father, an officer of the Soviet Army, her mother, a primary school teacher, and her brother Sergey, an officer of the Russian Army.

After secondary school, in 1974 Lyubov Vasilievna started to study at the Pediatric Department of the 2nd Pirogov Moscow State Medical University where she graduated with honors in 1980.

During 1980–1982 she studied in the clinical department of Infectious Diseases and in 1984 she was accepted into the laboratory of Microorganisms' Anatomy at N. F. Gamaleya Federal Research Center for Epidemiology & Microbiology, Moscow. Here Lyubov Vasilievna studied the ultrastructure of pathogens responsible for intestinal infections, and the features of blood during acute intestinal infections and persistence of bacteria. In 1987 she successfully defended her thesis "Clinical and diagnostic value of the study of morphological and functional state of peripheral blood of patients with typhoid fever by electron microscopy and morphometry" and continued her research on acute intestinal infections along with her studies on the role of bacteremia as a central link in the pathogenesis of all known bacterial infections.

Since 1995 Lyubov Vasilievna was Head of the Laboratory of Microorganisms' Anatomy of the N. F. Gamaleya Federal Research Center for Epidemiology & Microbiology. In 2001 Luybov Vasilievna defended her thesis "Bacteremia: a basis for the pathogenesis of infectious and non-infectious diseases" and became Doctor of Medical Sciences. This research pioneered numerous studies examining the interaction between the human body and microorganisms.

During her professional career Lyubov Vasilievna conducted researches in several directions: she investigated the interaction of different types of microorganisms in the environment and in clinical settings; the ultrastructure of *Helicobacter pylori* both *in vivo* and *in vitro* and the process of biofilm formation of *Salmonella typhimurium*, *Burkholderia cepacia*, *Legionella* spp., *Escherichia coli*, *Staphylococcus aureus*, and *Brevibacillus laterosporus*; she analyzed the structural features of biofilms in water systems and on medical supplies; she was dedicated to the research on coronaviruses in experimental pneumocystic pneumonia; she researched the infectious aspects of kidney and prostate stone formation and the biodegradation of dental prostheses under bacterial colonization. Moreover, Lyubov Vasilievna implemented her research with new methods of immunocytochemistry and preparation techniques, introduced a method for detecting circulating immune complexes from blood plasma for their ultrastructural analysis, and modified bacteriological procedures for isolating bacteria and their antigens from the patients' blood.

Lyubov Vasilievna made relevant contributions to the electron microscopy of microorganisms' anatomy both in the transmission and the scanning electron microscopies. She implemented this area of research devoting energies to correlate Focused Ion Beam and Electron Dispersive Spectroscopy with other microscopies. In 2009 the laboratory of Lyubov Vasilievna received one of the most modern electron and ion microscopes, which allowed the study of non-dehydrated biological samples with a significant reduction of artifacts.

An electron microscopy training course session at FEI Nanoport (Eindhoven, The Netherlands) in November 2009 was the starting point of an interesting and far-reaching scientific cooperation between Lyubov Vasilievna's laboratory team and an Italian team guided by Professor Marziale Milani of the University of Milano-Bicocca (Milan, Italy). A stable scientific connection developed a few months later and on February 2, 2011, a collaboration agreement between N. F. Gamaleya Federal Research Center for Epidemiology & Microbiology and the University of Milano-Bicocca was signed and the new work group officialized. The new team met not only in Milan and Moscow, but even in occasion of electron microscopy European meetings in order to discuss and present the scientific activity: thanks to the storytelling ability of Lyubov Vasilievna each time the debates were stimulating from a scientific point of view and refined from the human standpoint.

From these scientific jam sessions interesting ideas and future objects of investigation in the common research activity rose. The first topic of the research activity regarded *Bacillus subtilis* spores and crystals formation connected to their use as insecticides; the warhorse of the cooperative scientific activity was the biodestruction of plastic materials carried out by microorganisms, a scientific trend triggered by George Alexandrovich Avtandilov. This activity brought two important outcomes: on one hand the bacterial biodestructive action on organic and inorganic materials with consequent production of non-engineered nanoparticles, on the other hand the interaction of nanoparticles with the environment, the microorganisms, and the host organism (human body), with focus on the toxicological effects caused by nanosized materials. The results of the studies of Lyubov Vasilievna were presented in international journals and in several books, as well as at international congresses and conferences.

On December 8, 2015, the life of the outstanding scientist and dear friend Lyubov Vasilievna ended and she left her husband Alexander Georgievich Avtandilov and her son George Alexandrovich, along with numerous colleagues and friends.

Lyubov Vasilievna was not only a talented scientist and a good leader for her team, but also a great woman who took her group to heart balancing both the professional and the human aspects, and a true friend to many people, helpful and sincere. She used to describe her way of life with the motto of Claudius Aquaviva written in the manual for Jesuit spiritual directors *Industriae ad curandos animae morbos.*

Fortiter in re, suaviter in modo.
Powerfully in deed, pleasantly in manner.

She was a woman of culture, able to involve the listener in the field of her activity, framing the sometimes super-technical elements in complex stories resembling theatrical *pieces* narrated with theatrical times. She was able to look through a person understanding their soul. Her name, Lyubov, intended as *agape* (Greek: ἀγάπη), was a token of her disinterested and unconditional love for science, her team and friends, an attitude perfectly enclosed in the following quote.

> *I knew he was a good surgeon, but it was not the surgeon but the person I would stand in relation to, or, rather, the man in whom, I hoped, the surgeon and the person would be wholly fused.*
> *Oliver Sacks*

Her ability to feel the people, together with her formation as Medical Doctor and all the human and psychological connotations, made Lyubov Vasilievna the best leader for her diversified team: Svetlana Georgievna Andreevskaya, Tatiana Gennadievna Borovaya, Nina Sergeevna Glebova, Irina Victorovna Leushina, Tatiana Michailovna Poliakova, Natalia Vladimirovna Shevlyagina, Tatiana Naumovna Timoshina, Tatiana Alexandrovna Smirnova, and Elena Andreevna Kost who left us too early.

Lyubov Vasilievna Didenko will be remembered forever.

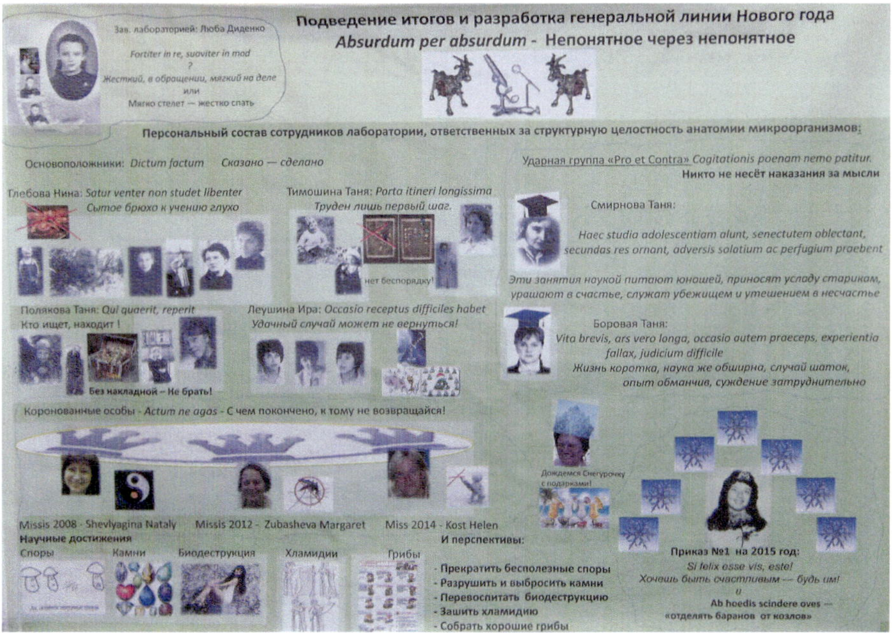

Picture of the poster in the meeting room at N. F. Gamaleya Federal Research Center for Epidemiology & Microbiology representing the staff of the Laboratory of Microorganisms' Anatomy with personal comments in Latin by Lyubov Vasilievna Didenko

Commemoration Day

On April 6, 2016, the laboratory team of Dr. L.V. Didenko organized a Commemoration Day at the N. F. Gamaleya Federal Research Center for Epidemiology & Microbiology, Moscow. Several scientists from other research institutes and hospitals were invited to remember Dr. L.V. Didenko and to talk about the scientific collaboration and the activities shared with her.

The research group of A.G. Avtandilov gave a speech about the effect of magnesium orotate on connective tissue and the cardiac inotropic function in patients with mitral valve prolapse. It is known that magnesium ions play an important role in the proper functioning of connective tissue and its matrix. Chronic deficiency of magnesium ions leads to disorders of the metabolism of the structural components of the connective tissue. The aim of the research was to study the morphological features of the connective tissue and the myocardial contractility in patients with mitral valve prolapse before and after treatment with magnesium orotate. It was shown that magnesium drugs have a positive impact on the connective frame and on the extracellular matrix, enhancing the formation of proteoglycans by fibroblasts, with a significant increment of the diffusion ability of the connective tissue stroma of the myocardium, and its elasticity and extensibility.

Moreover, the same team presented a study on the morphofunctional characteristics of fibroblasts (McCoy cell line) cultured with magnesium preparations. Fibroblasts are the main structural components of loose connective tissue, which is part of the human body organs that carries out support, trophic, and regulatory functions. The aim of the work was to study the effect of magnesium orotate, magnesium/pyridoxine combination, and magnesium sulfate on the morphofunctional characteristics of fibroblasts in cell cultures (McCoy line). Significant changes in the morphofunctional state of fibroblasts were noted when magnesium orotate or magnesium/pyridoxine combination was added in the medium; changes were absent when magnesium sulfate was added. Magnesium orotate and magnesium/pyridoxine combination promoted both the synthetic and secretory activity of fibroblasts. This was expressed in the formation of amorphous and fibrous components in well-formed cell layers. After treatment with magnesium orotate and magnesium/pyridoxine combination, in many areas the center of the matrix looked more massive. The matrix layer in formulations containing magnesium orotate was more pronounced and occupied a larger area. After treatment with magnesium sulfate there was an increase in the amount of fibroblasts compared to the control, but no significant morphological differences of the cells and the matrix layer were observed.

Magnesium orotate had a predominant influence on the biosynthetic processes of the cells, including the synthesis of amorphous and fibrous components (protocollagen and elastin) of the extracellular matrix, which in the clinical setting increases the diffusion ability of interstitial bodies.

The group of G.E. Stolyarenko using scanning and transmission electron microscopy identified the morphological features of epiretinal samples removed during vitrectomy in patients with lamellar macular hole or epiretinal membrane. Porous coral-like structures were discovered on the retinal side of the inner limiting membrane. The functional significance of these previously unknown structures and the effect of partial removal during surgery should be studied.

V.G. Zhukhovitsky presented his studies about the morphological characterization of clinical isolates of *Helicobacter pylori* and its interaction with the gastric mucosa. The investigation of bioptic samples of gastric mucosa from patients affected by chronic gastritis and duodenal ulcer using scanning electron microscopy showed the presence of significant amounts of bacteria on the mucosa surface. Bacteria were immersed in an amorphous layer and only a few of them were found on the surface of the amorphous substance. The analysis of biopsies' ultrathin sections showed that bacteria interacted with cells of the gastric mucosa through the bacterial cell wall; however phagosome was not formed. *Helicobacter pylori* cells came in contact with disrupted mucus-producing cells and penetrated into epithelial cells. Moreover, an infiltration by leukocytes and plasma cells in the mucosa was detected.

N.V. Shevlyagina together with T.G. Borovaya, A.M. Ivanova, and A.N. Narovlyansky presented studies on the structural changes in the ovaries of guinea pigs in the experimental model of *Herpes simplex* virus type 2 infection. The light microscopy analysis of the ovaries of guinea pigs showed lesions at follicles at different stages of folliculogenesis, the decrease in the number of follicles and their degenerative structure, the absence of Graafian (mature) follicles, and finally a cystic degeneration of the follicles. In the stromal ovarian department, the appearance of divergent strands of fibroblasts and a damaged ciliary apparatus of epithelial cells in the ducts of the ovary network (*rete ovarii*) were observed.

G.A. Avtandilov talked about the research on the biodegradation of polymer materials in dentistry. The aim of his study was to show the biodestruction of dental plastic material under the bacterial and fungal influence (*Staphylococcus aureus*, *Candida albicans*, *Pseudomonas aeruginosa*, *Enterococcus faecium*). It was found that bacteria and fungi form biofilms on the surface of dental plastic materials and destroy it. *Foci* of persistent microorganisms were localized in correspondence with surface defects (cracks and cavities) of denture samples. Moreover it was shown that the biofilm found on the surface of desquamated polymer particles could lead to dissemination of infectious agents causing dysbiosis in the oral cavity.

The team of S.A. Borzenok discussed the problem of corneal sample preparation for scanning electron microscopy. The study was conducted using eight cadaver eyes by four dead donors without signs of corneal pathology and not demanded for transplantation. The researcher formed the corneal flap using a femtosecond laser and applied different preparation procedures for scanning electron microscopy. The use of 4.0% formaldehyde and Ito-Karnovsky reagent as fixatives with subsequent alcohol dehydration and critical point drying could form artifacts (chips, crystal-like structures) on the surface. The optimal sample preparation method of cadaver human cornea for scanning electron microscopy resulted to be the fixation in 4.0% formaldehyde solution or in Ito-Karnovsky without subsequent dehydration with alcohol and critical point drying. This method preserved the morphological appearance of the cornea nearest to the physiological condition and made it possible to conduct a detailed assessment of the quality of the surface of the corneal stromal bed.

The group of V.M. Elinson presented studies about physico-chemical properties and resistance to biodegradation of fluorinated surfaces. The aim of the work was to study the surface energy and roughness parameters of nanostructured fluorinated surfaces based on polytetrafluoroethylene (PTFE), modified by ion-plasma methods, and to determine the materials biodegradation under the influence of *Staphylococcus aureus* cells. The electron microscopy studies of the interaction processes of *Staphylococcus aureus* with PTFE surfaces, formed by treatment of CF_4 ions and deposition of fluorocarbon film with the use of gas phase $CF_4 + C_6H_{12}$, has shown that, apparently, optimal combination of fluorine concentration and definite surface relief determine the coating resistance to biofouling and biodegradation. On the basis of measurements of contact wetting angle and calculations of the total specific surface energy, it was found that the formation of nanostructures, providing the absence of *Staphylococcus aureus* cells adhesion, was accompanied by a sharp drop of the surface energy.

The group of V.N. Kopyltsov discussed the ability of different types of *Staphylococci* to form biofilms and their effects on cell cultures. The studied bacteria were strains of *S. aureus*, *S. haemolyticus*, and *S. epidermidis* derived from neonatal autopsy materials. The visual tests of biofilm formation by each strain and testing of the impact of the strains on HT-29 cell culture were carried out by scanning electron microscopy.

The obtained data showed the toxic effects of all studied *Staphylococci* strains on HT-29 cells. Both *S. haemolyticus* and *S. epidermidis* formed biofilm faster than *S. aureus*.

V.V. Janowski and his team talked about the evaluation (by scanning electron microscopy) of shunts removed from children with exudative otitis media (EOM). On the surface of ventilation tubes favorable conditions were created for the development of fungal and/or bacterial biofilms. Hence the appropriate correction of drug therapy in the post-shunting of the tympanic cavity is required. It was found that a shunt which remains in the eardrum for more than 12 months is impractical because, apparently, relapse of EOM can be induced by the degradation of the shunt material. If aeration of the tympanum is necessary, the shunt must be changed.

The team of M. Milani talked about the electron microscopy study of the biodegradative action of *Staphylococcus aureus* on polymeric dental prostheses. This long-term investigation, conducted thanks to a Russian Federation-Italy prolific cooperation, brought two important outcomes related on one hand to the bacterial biodestructive action on organic and inorganic materials with consequent production of non-engineered nanoparticles, and on the other hand to the interaction of nanoparticles with the environment, the microorganisms, and the host organism (human body), with focus on toxicological effects caused by nanosized materials.

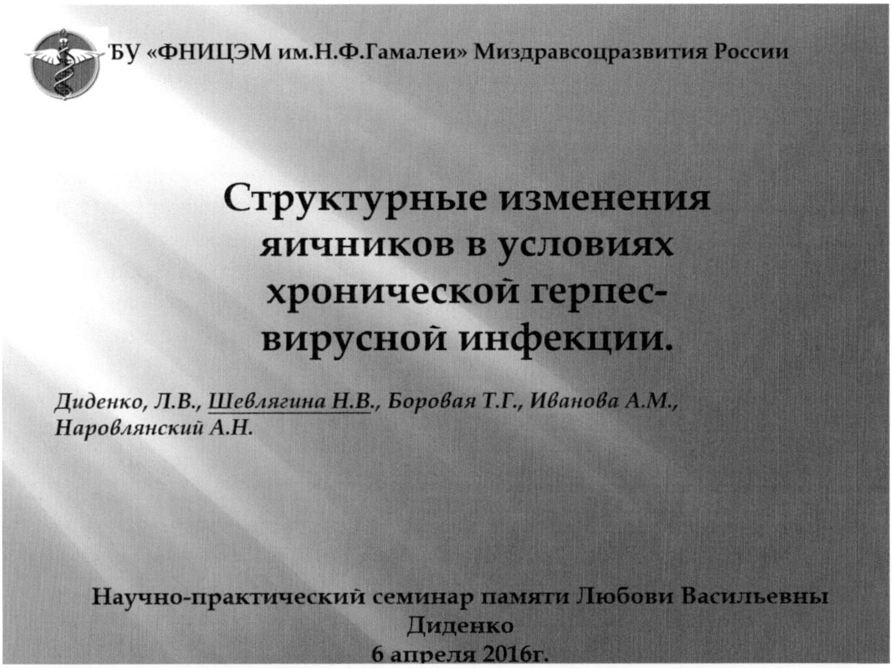

Opening slides of the oral presentations of Natalia Vladimirovna Shevlyagina, Marziale Milani, Roberta Curia, and Francesco Tatti

The team of A.L. Mulyukin gave a report about forms of long-term survival of *Pseudomonas*. Electron microscopy studies established a morphological heterogeneity in populations of *Pseudomonas* in macrocapsules and the presence of cells analogous to resting cyst-like forms. A much deeper insight on how bacteria persist in natural conditions and how potential forms of *P. aeruginosa* persist in the body during chronic infections is needed. Encapsulated cells can be recommended as components of detecting systems to test the action of antimicrobial drugs at fighting resistant infections and as the basis for stable bacterial preparations.

S.O. Navolnev talked about the development of a new computer program for quantifying microorganisms in digital images obtained by electron microscopy, since the computer programs available so far are extremely insufficient for such complex images.

V.N. Olesova and her team discussed the biodegradation and microbial colonization of ceramic and polymeric dental materials. An electron microscopy investigation was conducted on pressed ceramic and light-curing composite. Results confirmed that compared to composite materials, ceramic samples were significantly less colonized by microorganisms and were more resistant to biodegradation.

The team of Y.M. Romanova talked about a study on nonculturable forms of pathogenic bacteria. The discussion highlighted the comparative characteristics of noncultivated forms of bacteria and persistent cells, and a possible mechanism for their formation and existence in biofilms was reported.

The group of T.A. Smirnova discussed the biofilms of spore-forming bacteria. The electron microscopy analysis of *Bacillus subtilis* and *Brevibacillus laterosporus* biofilms showed differences in biofilms formed by different types of *Bacilli*. In the case of *B. subtilis* vegetative cells were free from defects, which were specific to *B. laterosporus*. The biofilms of the two species of *Bacilli* were similar in some respects. This study suggested that *B. laterosporus* and *B. subtilis* have a total capacity for sporulation and biofilm formation.

The team of E.R. Tolordava studied biofilms on the surface of kidney stones through electron microscopy and energy dispersive X-ray analyses. The results revealed that bacterial biofilms were present only in those areas where Magnesium, Nitrogen, and Phosphorous were simultaneously detected, and absent in those parts of the stones where only Calcium or Calcium and Phosphorous were found, thus highlighting the existence of a direct relationship between the chemical composition of urinary stones and biofilm formation of uropathogenic bacteria. The study concluded that microorganisms are capable of existing in biofilm on the stones of patients with urolithiasis, colonizing the urinary system, maintaining chronic infections and inflammation, and being the cause of recurrence of urolithiasis.

To conclude, this chapter is dedicated to Dr. L.V. Didenko; here we present a list of her publications.

Publications of Lyubov V. Didenko: 2007–2016

Столяренко ГЕ, Колчин АА, Диденко ЛВ, Боровая ТГ, Шевлягина НВ. Пористая коралловидная структура – новое представление о морфологии внутренней пограничной мембраны сетчатки? Вестник офтальмологии. 2016;132(6):70–7.

Елинсон ВМ, Диденко ЛВ, Шевлягина НВ, Автандилов ГА, Гайдарова АХ, Лямин АН, Исследование процессов колонизации *Staphylococcus aureus* наноструктурированных фторсодержащих поверхностей, сформированных разными методами ионно-плазменной технологии. Бюллетень экспериментальной биологии и медицины. 2016;162(7):84–7.

Корниенко МА, Копыльцов ВН, Шевлягина НВ, Диденко ЛВ, Любасовская ЛА, Припутневич ТВ, Ильина ЕН. Способность стафилококков различных видов к образованию биопленок и их воздействие на клетки человека. Молекулярная генетика, микробиология и вирусология. 2016;1:20–7.

Milani M, Didenko LV, Avtandilov GA, Curia R, Erega A, Shevlyagina NV. Electron Microscopy Documents the Microorganisms' Biodestructive Action on Polyurethane and the Production, Internalization and Vesicular Trafficking of Nanoparticles. Br J Appl Sci Techno. 2016;12(3).

Диденко ЛВ, Автандилов ГА, Ипполитов ЕВ, Царева ЕВ, Смирнова ТА, Шевлягина НВ, Царев ВН. Формирование биопленок на стоматологических полимерных материалах как основа персистенции микроорганизмов при патологии зубов и пародонта. Эндодонтия Today. 2015;4.

Боровая ТГ, Диденко ЛВ, Наровлянский АН, Шевлягина НВ, Иванова АМ, Санин АВ, Пронин АВ. Гистологические особенности сети яичника в остром периоде генитальной герпесвирусной инфекции. Морфологические ведомости. 2015;23(2):15–20.

Ипполитов ЕВ, Диденко ЛВ, Царев ВН. Особенности морфологии биопленки пародонта при воспалительных заболеваниях десен (хронический катаральный гингивит, хронический пародонтит, кандида-ассоциированный пародонтит) по данным электронной микроскопии. Клиническая лабораторная диагностика. 2015;60(12):59–64.

Диденко ЛВ, Автандилов ГА, Смирнова ТА, Шевлягина НВ, Царев ВН, Лебеденко ИЮ, Елинсон ВМ, Тиганова ИГ, Романова ЮМ, Ипполитов ОВ. Исследование процессов колонизации и персистенции микроорганизмов на искусственных материалах медицинского назначения. Журнал микробиологии, эпидемиологии и иммунобиологии. 2015;5:64–9.

Олесов ЕЕ, Лесняк АВ, Узунян НА, Диденко ЛВ, Автандилов ГА, Юффа ЕП, Адамчик АА. Экспериментальное изучение биодеградации и микробной колонизации реставрационных стоматологических материалов. Российский стоматологический журнал. 2015;19(4):4–6.

Наровлянский АН, Иванова АМ, Шевлягина НВ, Диденко ЛВ, Боровая ТГ, Изместьева АВ, Санин АВ, Пронин АВ, Ершов ФИ. Эффективность применения полипренилфосфатов в экспериментальной модели генитального герпеса. Voprosy virusologii. 2015;60(4):9–13.

Автандилов АГ, Воронов ИА, Лебеденко ИЮ, Диденко ЛВ, Смирнова ТА, Шевлягина НВ, Стафилококки в ротовой полости и их роль в биодеструкции съемных неметаллических протезов. Российский стоматологический журнал. 2015;19(1):14–20.

Romanova YuM, Mulabaev NS, Tolordava ER, Seregin AV, Seregin IV, Alexeeva NV, Stepanova TV, Levina GA, Barkhatova OI, Gamova NA, Goncharova SA, Didenko LV, Rakovskaya IV. Microbial communities on kidney stones. Mol Gen Microbiol Virol. 2015;30(2):78–84.

Пушкарева ВИ, Диденко ЛВ, Овод АА, Ермолаева СА. Взаимодействие *Escherichia coli* с растениями на популяционном и клеточном уровнях. Успехи современной биологии. 2015;135(3):297–306.

Azizbekyan RR, Didenko LV, Smirnova TA, Kuzin AI, Shevlyagina NV, Zubasheva MV, Nikolaenko MA. Biofilm formation and sporulation in *Brevibacillus laterosporus*. Appl Biochem Microbiol. 2015;51(9):866–72.

Диденко ЛВ, Боровая ТГ, Кост ЕА, Улубиева ЕА, Тимофеева АВ, Автандилов АГ, Морфофункциональная характеристика фибробластов клеточной культуры McCoy при культивировании с препаратами магния. Рациональная фармакотерапия в кардиологии. 2015;11(2):170–6.

Диденко ЛВ, Кардаш ГГ, Смирнова ТА, Толордава ЭР, Зубашева МВ, Куршин ДА, Емшанов ОВ, Изучение механизма действия третичных алкиламинов на клинических изолятах *Escherichia coli* и *Staphylococcus aureus*. Дезинфекционное дело. 2015;91(1):32–6.

Терентьева НА, Тимченко НФ, Балабанова ЛА, Голотин ВА, Белик АА, Бакунина ИЮ, Диденко ЛВ, Рассказов ВА. Влияние ферментов на формирование бактериальных биопленок. Здоровье. Медицинская экология. Наука. 2015;60(2):86–94.

Curia R, Milani M, Didenko LV, Avtandilov GA, Shevlyagina NV, Smirnova TA. Beyond the biodestruction of polyurethane: *S. aureus* uptake of nanoparticles is a challenge for toxicology. In Méndez-Vilas A, editor. Microscopy: advances in scientific research and education. Formatex Research Center; 2014. p. 16–23.

Didenko LV, Shevlyagina NV, Curia R, Erega A, Milani M. Electron microscopy furthers the investigation of bacteria-nanoparticles interactions subcellular dynamics. Global J Med Res D Radiol Diagn Imag Instrument. 2014;14(4):13–23.

Didenko LV, Tolordava ER, Perpanova TS, Shevlyagina NV, Borovaya TG, Romanova YuM, Cazzaniga M, Curia R, Milani M, Savoia C, Tatti F. Electron microscopy investigation of urine stones suggests how to prevent post-operation septic complications in nephrolithiasis. J Appl Med Sci. 2014;3(4):19–34.

Виноградов ИВ, Ким ДГ, Диденко ЛВ. Первый опыт комплексного электронно-микроскопического и бактериологического исследования камней предстательной железы. Андрология и генитальная хирургия. 2014;15(3):87–91.

Воронов ИА, Митрофанов ЕА, Калинин АЛ, Семакин СБ, Диденко ЛВ, Автандилов ГА. Разработка нового покрытия из карбида кремния для защиты зубных протезов от биодеструкции. Российский стоматологический журнал. 2014;1:4–9.

Voronina OL, Kunda MS, Aksenova EI, Ryzhova NN, Semenov AN, Petrov EM, Didenko LV, Lunin VG, Ananyina YV, Gintsburg AL. The characteristics of ubiquitous and unique *Leptospira* strains from the collection of Russian Centre for Leptospirosis. BioMed Res Int. 2014;Article ID 649034.

Диденко ЛВ, Иванов АИ, Смирнова ТА, Толордава ЭР, Зубашева МВ, Кардаш ГГ, Куршин ДА, Емшанов ОВ, Автандилов ГА. Действие третичных алкиламинов на планктонные культуры и биопленки *Escherichia coli* и *Staphylococcus aureus*. Farmatsiya. 2014;7:44–9.

Диденко ЛВ, Смирнова ТА, Толордава ЭР, Зубашева МВ, Кардаш ГГ, Куршин ДА, Емшанов ОВ, Автандилов ГА. Влияние третичных алкиламинов на биопленки, образованные *Escherichia coli* и *Staphylococcus aureus* (бактериологическое и электронно-микроскопическое исследование). Дезинфекционное дело. 2014;2:40–5.

Curia R, Milani M, Didenko LV, Shevlyagina NV. Electron microscopy broadens the horizons of toxicology: the role of nanoparticles vehiculated by bacteria. Curr Top Toxicol. 2013;9:93–8.

Didenko L, Avtandilov G, Shevlyagina N, Shustrova N, Smirnova T, Lebedenko I, Curia R, Savoia C, Tatti F, Milani M. Nanoparticles production and inclusion in *S. aureus* incubated with polyurethane: an electron microscopy analysis. Open J Med Imag. 2013;3(2):69–73.

Zigangirova NA, Zigangirova NA, Rumyantseva YP, Morgunova EY, Kapotina LN, Didenko LV, Kost EA, Koroleva EA, Bashmakov YK, Petyaev IM. Detection of *C. trachomatis* in the serum of the patients with urogenital chlamydiosis. BioMed Res Int. 2013:Article ID 489489.

Сахарова АВ, Диденко ЛВ, Муравина ТИ, Чайковская РП, Кост ЕА, Мир-Касимов МФ. Прямая морфологическая детекция *Borrelia burgdorferi* в мышечных биоптатах: возможность связи нервно-мышечной патологии с боррелиозом. Нервно-мышечные болезни. 2013;1:35–46.

Годова ГВ, Овод АА, Калашникова ЕА, Князев АН, Пушкарёва ВИ, Диденко ЛВ, Ермолаева СА, Образование биопленок *Listeria monocytogenes* при взаимодействии с растительными клетками. XXI век: итоги прошлого и проблемы настоящего плюс. 2013;2(9):62–9.

Боровая ТГ, Шевлягина НВ, Иванова АМ, Наровлянский АН, Калмыкова НВ, Третьяков ОВ, Диденко ЛВ. Структурные изменения коры надпочечников при экспериментальной генитальной герпесвирусной инфекции. Arkhiv Anatomii, Gistologii i Embriologii. 2013;144(6):52–7.

Годова ГВ, Пушкарева ВИ, Калашникова ЕА, Овод АА, Диденко ЛВ, Князев АН, Ермолаева СА. Формирование биопленок *Listeria monocytogenes* при взаимодействии с клетками овощных культур. Известия Тимирязевской сельскохозяйственной академии. 2013;5:50–9.

Пушкарева ВИ, Диденко ЛВ, Годова ГВ, Овод АА, Калашникова ЕА, Ермолаева СА. *Listeria monocytogenes* – взаимодействие с агрокультурами и стадии формирования биопленки. Эпидемиология и вакцинопрофилактика. 2013;1:42–9.

Автандилов АГ, Дзеранова КМ, Боровая ТГ, Диденко ЛВ. Влияние оротата магния на соединительнотканный каркас и инотропную функцию сердца у пациентов с пролапсом митрального клапана. Клинико-морфологическое исследование. Рациональная фармакотерапия в кардиологии. 2013;9(4).

Хон ВЭ, Загородний НВ, Комлев ВС, Фадеева ИВ, Булгаков ВГ, Сергеева НС, Свиридова ИК, Толордава ЭР, Диденко ЛВ. Влияние степени замещения кальция серебром в трикальцийфосфате на его биологические свойства *in vitro*. Вестник травматологии и ортопедии им. Н.Н. Приорова. 2013;4:23–8.

Didenko LV, Avtandilov GA, Shevlyagina NV, Smirnova TA, Lebedenko IY, Tatti F, Savoia C, Evans G, Milani M. Biodestruction of polyurethane by *Staphylococcus aureus* (an investigation by SEM, TEM and FIB). In Méndez-Vilas A, editor. Current microscopy contributions to advances in science and technology. Formatex Research Center; 2012. p. 323–34.

Зигангирова НА, Заякин ЕС, Капотина ЛН, Кост ЕА, Диденко ЛВ, Давыдова ДЮ, Румянцева ЮП, Гинцбург АЛ. Создание ингибиторов системы секреции типа III C. Trachomatis, подавляющих развитие острой и хронической хламидийной инфекции. *Acta Naturae* (русскоязычная версия). 2012;4(2):90–101.

Дронина ЮЕ, Карпова ТИ, Садретдинова ОВ, Диденко ЛВ, Тартаковский ИС. Особенности формирования биопленок легионелл в искусственных и природных водных системах. Zhurnal mikrobiologii epidemiologii i immunobiologii. 2012;4:76–80.

Диденко ЛВ, Садретдинова ОВ, Шевлягина НВ, Автандилов ГА, Новокшонова ИВ, Карпова ТИ, Груздева ОА, Тартаковский ИС. Морфологические особенности биопленок в потенциально опасных водных системах. Эпидемиология и инфекционные болезни. 2012;1:15–20.

Громов АВ, Никитин КЕ, Карпова ТА, Бартов МС, Мишина ДМ, Субботина МЕ, Шевлягина НВ, Сергиенков МА, Соболева ЛА, Котнова АП, Шарапова НЕ, Семихин АС, Диденко ЛВ, Карягина АС, Лунин ВГ. Разработка методики получения деминерализованного костного матрикса с максимальным остаточным содержанием нативных факторов роста костной ткани. Биотехнология. 2012;5:66–75.

Диденко ЛВ, Перепанова ТС, Толордава ЭР, Боровая ТГ, Шевлягина НВ, Эгамбердиев ДК, Голованов СА, Романова ЮМ. К вопросу об инфекционном генезе камней почек (электронно-микроскопическое исследование). Урология. 2012;3:4–7.

Борзенок СА, Онищенко НА, Тонаева ХД, Комах ЮА, Сускова ВС, Сусков СИ, Диденко ЛВ, Шевлягина НВ, Кост ЕА. Длительная нормотермическая консервация лимбальных трансплантатов как способ повышения количества и активности ММСК-подобных лимбальных клеток. Вестник трансплантологии и искусственных органов. 2012;14(2):81–9.

Толордава ЭР, Тиганова ИГ, Алексеева НВ, Степанова ТВ, Терехов АА, Егамбердиев ДК, Мулабаев НС, Шевлягина НВ, Диденко ЛВ, Романова ЮМ. Микрофлора почечных камней при мочекаменной болезни и поиск средств борьбы с биопленками уропатогенных бактерий. Zhurnal mikrobiologii epidemiologii i immunobiologii. 2012;4:56–62.

Романова ЮМ, Диденко ЛВ, Толордава ЭР, Гинцбург АЛ. Биопленки патогенных бактерий и их роль в хронизации инфекционного процесса. Поиск средств борьбы с биопленками. Vestnik Rossiiskoi Akademii Meditsinskikh Nauk/Rossiiskaia Akademiia Meditsinskikh Nauk. 2011;10:31–9.

Romanova YM, Tiganova IG, Khmel IA, Alekseeva NV, Stepanova TV, Shevlyagina NV, Didenko LV, Goryachev SN, Ginzburg AL. Biofilms of *Burkholderia cepacia*: characteristics of mutants with modified formation capacity. Mol Gen Microbiol Virol. 2011;26(3):93–101.

Тахчиди ХП, Шкворченко ДО, Шарафетдинов ИХ, Шацких АВ, Новиков СВ, Норман КС, Бибилашвили РШ, Скрыпина НА, Диденко ЛВ, Шевлягина НВ. Экспериментальное обоснование использования миниплазмина с целью индукции задней отслойки стекловидного тела. Вестник Оренбургского государственного университета. 2011;14(133):363–6.

Тахчиди ХП, Шкворченко ДО, Шарафетдинов ИХ, Новиков СВ, Шацких АВ, Норман КС, Бибилашвили РШ, Скрыпина НА, Диденко ЛВ, Шевлягина НВ, Крикун ЕН. Экспериментальное обоснование использования миниплазмина с целью индукции задней отслойки стекловидного тела. Научные ведомости Белгородского государственного университета. Серия: Медицина. Фармация. 2011;16(15):194–8.

Пашко ЮП, Зигангирова НА, Капотина ЛН, Моргунова ЕЮ, Колкова НИ, Диденко ЛВ. Особенности распространения в организме *Chlamydia trachomatis* при хроническом течении урогенитального хламидиоза и детекция возбудителя в сыворотке крови. Фундаментальные исследования. 2010;7:50–7.

Навольнев СО, Диденко ЛВ. Программа для определения морфологических особенностей биологических микрообъектов на изображениях полученных от микроскопа. Медицинский алфавит. 2010;2(1):41–2.

Боровая ТГ, Шевлягина НВ, Диденко ЛВ. Интраовариальные регуляторы фолликулогенеза. Uspekhi Fiziologicheskikh Nauk. 2010; 41(1):58–74.

Petyaev IM, Zigangirova NA, Petyaev AM, Pashko UP, Didenko LV, Morgunova EU, Bashmakov YK. Isolation of *Chlamydia pneumoniae* from serum samples of the patients with acute coronary syndrome. Int J Med Sci. 2010;7(4):181–190.

Smirnova TA, Didenko LV, Azizbekyan RR, Romanova YuM. Structural and functional characteristics of bacterial biofilms. Microbiology. 2010;79(4):413–23.

Smirnova TA, Didenko LV, Tiganova IG, Andreevskaya SG, Alekseeva NV, Stepanova TV, Romanova YuM. Study of the structures of biofilms formed by *Salmonella typhimurium* bacteria on abiotic surfaces by the methods of light and transmission electron microscopy. Appl Biochem Microbiol. 2010;46(7):706–11.

Туманян ЭР, Иванова ЕС, Любимова ТС, Субхангулова ЭА, Боровая ТГ, Диденко ЛВ, Шевлягина НВ. Морфологическая оценка изменений трабекулярной сети угла передней камеры глаза после селективной лазерной трабекулопластики по *M. Latina* и селективной лазерной активации трабекулы. Офтальмохирургия. 2010;3:39–42.

Боровая ТГ, Диденко ЛВ, Шевлягина НВ. Особенности овариального фолликулогенеза у морских свинок в модели рецидивирующего генитального герпеса. Arkhiv Anatomii, Gistologii i Embriologii. 2009;136(4).

Диденко ЛВ, Андреевская СГ, Тиганова ИГ, Смирнова ТА, Романова ЮМ. Роль сальмонелл в билиарном литогенезе. Экспериментальная и клиническая гастроэнтерология. 2009;3.

Timofeeva LM, Kleshcheva NA, Moroz AF, Didenko LV. Secondary and tertiary polydiallylammonium salts: novel polymers with high antimicrobial activity. Biomacromolecules. 2009;10(11):2976–86.

Пашко ЮП, Зигангирова НА, Петяев ЮМ, Капотина ЛН, Моргунова ЕЮ, Колкова НИ, Диденко ЛВ, Юдина ТИ. Современные аспекты диагностики хронического хламидиоза, вызванного персистирующими формами хламидий. Zhurnal mikrobiologii epidemiologii i immunobiologii. 2009;4:89–93.

Shchelkunov IS, Shchelkunova TI, Shchelkunov AI, Kolbassova YP, Didenko LV, Bykovsky AP, First detection of a viral agent causing disease in farmed sturgeon in Russia. Diseases of Aquatic Organisms. 2009;86(3):193–203.

Шевлягина НВ, Иванова АМ, Боровая ТГ, Диденко ЛВ, Березина ЛК, Наровлянский АН, Ершов ФИ. Морфофункциональные изменения яичников морских свинок в экспериментальной модели хронической герпес-вирусной инфекции. Byulleten’ Eksperimental’noi Biologii i Meditsiny. 2008;146(8):221–5.

Диденко ЛВ, Быковский АФ, Каражас НВ, Дрынов ИД, Константинова НД, Рыбалкина ТН, Корниенко МН. Обнаружение коронавируса при экспериментальной пневмоцистной инфекции. Meditsinskaya Parazitologiya i Parazitarnye Bolezni. 2008;2: 8–11.

Навольнев СО, Диденко ЛВ, Андреевская СГ. Программа для выявления новых объектов на цифровых изображениях и применение ее для анализа микроскопического строения биопленок. Klinichescheskaya Laboratornaya Diagnostika. 2008;9.

Шевлягина НВ, Иванова АМ, Боровая ТГ, Диденко ЛВ, Березина ЛК, Наровлянский АН. Морфологические особенности яичников в условиях действия вируса простого герпеса 2-го типа. Фундаментальные исследования.2008;4.

Шевлягина НВ, Иванова АМ, Боровая ТГ, Диденко ЛВ, Березина ЛК, Наровлянский АН. Изменения овариального фолликулогенеза при экспериментальной модели герпес-вирусной инфекции у морских свинок. Успехи современного естествознания. 2008; 2.

Smirnova TA, Didenko LV, Andreev AL, Alekseeva NV, Stepanova TV, Romanova YuM. Electron microscopic study of *Burkholderia cepacia* biofilms. Microbiology. 2008;77(1):55–61.

Шевлягина НВ, Боровая ТГ, Диденко ЛВ, Наровлянский АН, Иванова АМ. Реактивные изменения фолликулярных гистионов яичников морских свинок в условиях экспериментального моделирования генитальной формы хронической герпес-вирусной инфекции. Фундаментальные исследования. 2007;12(2):293–4.

Зигангирова НА, Петяев ИМ, Пашко ЮП, Моргунова ЕЮ, Капотина ЛН, Диденко ЛВ, Юдина ТИ, Шубин СВ, Джикидзе ЭК, Аршба ИМ, Гинцбург АЛ. Генерализация инфекции у больных с урогенитальным хламидиозом. Клиническая микробиология и антимикробная химиотерапия. 2007;9(4):351–60.

Щелкунов ИС, Щелкунова ТИ, Щелкунов АИ, Колбасова ЮП, Диденко ЛВ, Быковский АФ, Герпесвирусная болезнь осетровых рыб в России. Российский ветеринарный журнал. Сельскохозяйственные животные. 2007;1:10–12.

Contents

Part I

Characters and Context

One of the aims of this book is to investigate how *Staphylococcus aureus*, able to colonize clinical prostheses, can produce polyurethane nanoparticles and how this process, typical also of other microorganisms, can interlace with the external world. This first part of the book is dedicated to present the main characters of the story and to discuss about an effective and far-reaching approach toward the nanoworld dynamics.

A general non-exhaustive description of *Staphylococcus aureus* is given with an excursus about its discovery, general features, and mechanisms of pathogenicity. Attention is devoted to the *Staphylococcus aureus* biofilm formation ability and biodegradative activity, and to the description of the production and absorption of polyurethane nanoparticles.

The second chapter discusses the history and the development of electron microscopes giving a general overview about the machines and the physics of electron microscopy, with focus on life sciences applications.

In the following chapter, the world of nanoobjects is featured starting from the description of nanotechnology and a classification of the properties and applications of nanosized materials. The overall knowledge about nanomaterials is presented in relation to the real-world knowledge and is paradoxically connected to the so-called microscopic world with its quantum laws.

In this top-down approach, microorganisms pervade that portion of the world external to them that is characterized by micrometric and nanometric sizes; the analysis of these processes requires the specific physics tools. In particular, the biodegradative action of *Staphylococcus aureus* on polyurethane prostheses, the production of nanoparticles and their subsequent internalization in surrounding bacteria and eukaryotic cells raise the main role of the bacterium from an infectious pathogen able to communicate and evade the immune system, to a vehicle of nanoparticles into host eukaryotic cells. *Staphylococcus aureus* contributes to extend in time and space the nanoparticles' story, sometimes in a bottom-up direction, increasing both the positive and negative potentialities of nanoparticles.

Staphylococcus aureus

Often I meditated on the subject and became the more convinced that there was a single cause and that the cause was some special germ.

Alexander Ogston

History

Microorganisms in the genus *Staphylococcus* were discovered in 1870 by Sir Alexander Ogston who examined via optical microscopy the pus evacuated by a leg lesion of a young man. After staining the sample with aniline violet solution, he observed a great number of "tangles, tufts, and chains of round organisms" together with pus cells and debris. The examination of hundreds of samples of pus led Ogston to claim that groups or chains of cocci (round microorganisms) were present in all cases of acute suppuration, i.e., the process of formation and discharging of pus. Sir Ogston differentiated two kinds of cocci: those in chains which corresponded to *Streptococci* (στρεπτόκοκκος from Greek streptos, chain, and kokkos, berry) as described by C. A. Theodor Billroth in 1874, and the ones in groups which Sir Ogston called *Staphylococci* (σταφυλόκοκκο from Greek staphyle, bunch of grapes, and kokkos, berry) (Etymologia: Streptococcus 2016; Licitra 2013). He also observed for the first time the presence of other forms of microorganisms, describing *Bacilli* and *Spirochetes* (Ogston and Witte 1984; GS 1965).

To investigate the role of *Staphylococci*, Sir Ogston conducted a series of experiments inoculating pus containing microorganisms in mice and managed to demonstrate that the inoculation induced into the animals the formation of abscesses (swollen areas within body tissue, containing an accumulation of pus) in which the number of microorganisms was greatly increased, meaning *Staphylococci* were alive. Then he proceeded inoculating cultured *Staphylococci* into mice and even in this case the formation of abscesses from which microbes were recoverable was observed. Sir Ogston's conclusions on *Staphylococci* stated their ability to at first start their proliferation locally, then to invade peripheral areas, and eventually gain

M. Milani et al., *Bacterial Degradation of Organic and Inorganic Materials*,
https://doi.org/10.1007/978-3-031-26949-3_1

access to the bloodstream, diffusing in the body and provoking septicemia (Cheng et al. 2011; Ogston and Witte 1984; GS 1965).

In those years where the scientific scene, especially in the field of microbiology, was held by German and French researchers, Sir Ogston was the only Briton to distinguish himself for his discoveries; despite this his studies were deeply criticized in his country (GS 1965).

In 1884 the German surgeon Anton F. J. Rosenbach isolated two *Staphylococcus* strains which he named after the different pigmentation of their colonies: *Staphylococcus aureus* (from Latin *Aurum* = Gold) and *S. albus* (from Latin *albus* = white), which was later called *S. epidermidis* due to its location on the skin (Licitra 2013).

General Characteristics

Staphylococcus aureus, one of the more than 36 species in the genus *Staphylococcus*, is a Gram-positive spherical bacterium of about 0.5–1.5 μm (Fig. 1.1) of diameter which usually grows in grape-like clusters. Its cells' division occurs on three orthogonal planes; sister cells, though able to change position, remain attached to each other forming the typical cluster (Cheng et al. 2011; Harris et al. 2002; Foster 1996; Tzagoloff and Novick 1977).

Staphylococcus aureus is a non-motile and non-spore forming bacterium able to grow between 15° and 45 ° C, with an optimum at 30–37 ° C, and able to tolerate salt concentration in the growth medium as high as 15%; it is a facultative anaerobe that can grow either by aerobic respiration or by fermentation.

Staphylococcus aureus is catalase positive (it can convert hydrogen peroxide in water and oxygen), oxidase negative, and nearly all strains of *Staphylococcus*

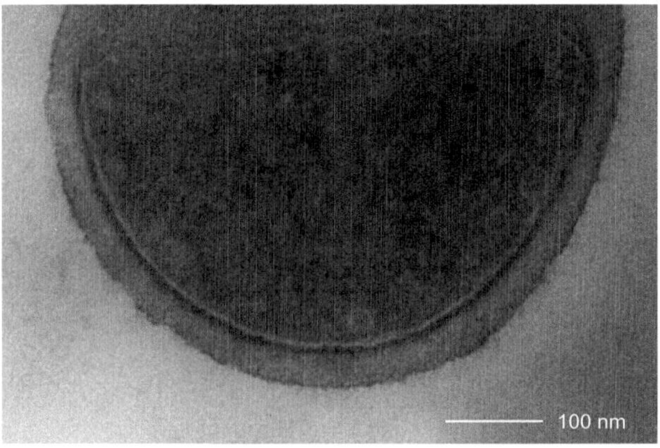

Fig. 1.1 TEM image of a portion of a *Staphylococcus aureus* cell from a control sample used as reference in the investigation of the biodestruction processes operated by *Staphylococcus aureus* cells on polyurethane

aureus are coagulase positive, an ability that allow bacterial cells to clot plasma to protect themselves in the first phases of infection (Harris et al. 2002; Foster 1996).

Normally *Staphylococci* spp. colonize warm-blooded animals and can be transmitted from one species to another either through aerosol or direct contact among infected individuals (Rossi et al. 2016). In human beings *Staphylococcus aureus* is a common commensal that can be found on the skin, in mucous membranes (nares, mouth, and throat), as well as in other body regions such as the intestinal tract and the mammary glands. The best habitat for *Staphylococcus aureus* is the squamous epithelium of the nares which provide a warm and moist environment. It is estimated that *Staphylococcus aureus* cells constantly colonize 20% of the population, the 60% is composed of intermittent carriers of the microbe, and the remaining 20% never carries the bacterium. Preadolescents usually carry a higher number of bacterial cells which result more dispersed all over the children's body than in adults. Unfortunately, the factors that promote the colonization of human surfaces, the mechanisms that determine a person's tendency to bacterial colonization, and the elements that allow some individuals to never carry *Staphylococcus aureus* are still not completely understood. Surely it is verified that the risk of contracting recurrent infections is increased in carriers and that healthcare-associated infections are commonly caused by the strains carried in the patients' nares; this is why some attempts to fight staphylococcal infections using topical antimicrobials directly on the nares were made with some success (Rasigade and Vandenesch 2014; Foster 2004; Lowy 2003).

Pathogenicity

Staphylococcus aureus is a common commensal that colonizes the human body and rapidly spreads in the population; usually its presence is asymptomatic but in some cases it can cause severe infections, thus becoming a pathogen. Though of high relevance, it is still not known whether genotypic differences exist between harmless *Staphylococcus aureus* isolates and infectious ones (Rasigade and Vandenesch 2014).

Natural external barriers as the skin provide protection from bacterial invasion and subsequent infections. If bacteria manage to breach the epithelium, the first line of defense is represented by neutrophils, leukocytes that ingest microbes and activate antimicrobial mechanisms until finally undergoing apoptosis to keep the inflammation under control. Unfortunately, *Staphylococcus aureus* has developed the ability to evade neutrophil killing and has turned from an extracellular pathogen to an intracellular one developing abilities to both replicate in phagosomes and induce anti-apoptotic mechanisms to survive in host cells. Moreover the bacterial colonization of the nares could operate as a reservoir for recurrent infections in compromised patients. Hence for patients at risk, carriage is considered very harmful; on the other hand it has been demonstrated that bacterial carriage sometimes provides some immunological benefits for the host as it was seen in patients that registered a lower risk of death after having developed *Staphylococcus aureus* bacteremia (Guerra et al. 2017; Rasigade and Vandenesch 2014; Fraunholz and Sinha 2012).

Staphylococcal infections that interest the skin consist of skin lesions, infections of hair follicles, acne, sties, and abscesses; others interest the middle ear and urinary tract. Though these infections are non-fatal, *Staphylococcus aureus* is considered by the World Health Organization one of the extremely dangerous pathogens that can cause serious infections, at time even deadly (in the United States of America each year it is estimated that 100,000 people die from *Staphylococcus aureus* infections), like pneumonia, meningitis, blood infections, infective arthritis, osteomyelitis, and endocarditis. *Staphylococcus aureus* is also responsible for colonizing implanted medical devices as prostheses, artificial heart valves, and catheters causing life-threatening bacteremia. Usually if only superficial tissue is interested by the infection, it is cured using pharmacological treatments, but in some cases when the infection is deeper surgery is required. *Staphylococcus aureus* is also the most common cause of surgical wounds contaminations and more and more infections are provoked by multiresistant strains (Hrebík et al. 2019). Some staphylococcal infections are also triggered by other microbes, such as viruses. It has been shown that interkingdom interactions between viruses and bacteria provoke coinfections that play a critical role in pathogenesis. The synergy between viruses and bacteria enhances immune modulation and tissue-remodeling mechanisms intervening on neutrophil cell death program in a sort of reciprocal communication or dialogue, and, as in the case of coinfection by influenza A virus and *Staphylococcus aureus*, the virus interacts directly with the bacterial surface increasing the bacterial adherence to respiratory epithelial cells (Rowe et al. 2019; Guerra et al. 2017).

Nowadays infections are treated with antibiotics bypassing the phage therapy previously adopted in the Soviet medicine, where a class of viruses called bacteriophages were used to infect bacteria and induce their death (Myelnikov 2018); however the infections provoked by the dangerously increasing number of antibiotic-resistant bacteria require the use of different and more powerful antibiotics to be contrasted. This also implies the prolongation of the time that patients spend in hospitals, with the risk to be exposed to new microbes that in a lot of cases leads to the death of the patient.

The pattern of antibiotic resistance, which at the beginning emerges in hospitals and then is spread to the community, unfortunately has been recurring through the years with every new drug available.

Prior to the introduction of antibiotics, the mortality of patients affected by *Staphylococcus aureus* bacteremia was over 80%. At the beginning of the 1940s penicillin improved the conditions of patients infected with this bacterium so that it was considered a miracle drug, but already in 1942 the first penicillin-resistant staphylococci were detected. At the end of the 1960s more than 80% of both community- and hospital-acquired staphylococcal isolates were resistant to penicillin and nowadays about the 90% of *Staphylococcus aureus* strains are resistant to penicillin. To overcome the situation other antibiotic derivatives of penicillin were developed. The introduction in 1961 of methicillin, the first of the semisynthetic penicillinase-resistant penicillins, was promptly followed by reports of methicillin-resistant *Staphylococcus aureus* (MRSA) strains. The first appearance of MRSA infections occurred in hospitals, but rapidly they were registered in subjects with no prior hospital exposure. MRSA infections were associated to a high mortality rate because as the penicillin-resistant strains also the MRSA isolates were resistant against other antimicrobial drugs. It is now known that almost 50% of *Staphylococcus aureus* strains are resistant to all β-lactam agents.

Fluoroquinolones were introduced in the 1980s for the treatment of infections caused by Gram-negative bacteria, but due to their spectrum they were also adopted to treat infections caused by Gram-positive pneumococci and staphylococci. Even in this case resistance against quinolones emerged relatively quickly in *Staphylococcus aureus* especially among the MRSA strains. Moreover it was observed that the use of quinolones for the treatment of infections caused by other pathogens in subjects colonized with *Staphylococcus aureus* (on the skin, in the nares, and on mucosal surfaces) exposed the patients to subtherapeutic concentrations of antibiotic, thus increasing the risk of becoming colonized with a reservoir of resistant mutants.

In the last years patients affected with MRSA strains were treated with vancomycin, but also vancomycin-resistant bacteria were developing. In 1997 a case of vancomycin intermediate-resistant *Staphylococcus aureus* (VISA) was reported for the first time. Since then, cases of vancomycin-resistant *Staphylococcus aureus* (VRSA) have been emerging, generating great concern in the medical communities (Lowy 2003).

At the present time the antimicrobial drugs effective against resistant strains comprehend quinupristin-dalfopristin, linezolid, and daptomycin. Due to the extended overuse and misuse of antibiotics by people, resistance develops extremely fast among bacterial strains, diminishing the efficacy of the currently available antimicrobial agents; hence the need for the development of new drugs and novel mechanisms of action to target bacteria is strong (Lowy 2003). Probably due to the need to find different approaches, recent interest sparked again for the previously adopted phage therapy (Hassan et al. 2021; Myelnikov 2018).

Staphylococcus aureus does not cause only infections, the bacterium also expresses surface proteins called microbial surface components recognizing adhesive matrix molecules (MSCRAMMs) that help the pathogen to attach to host tissue, enhancing its colonization (Plata et al. 2009). In addition *Staphylococcus aureus* secretes a series of virulence factors, mostly toxins, that help in the evasion of host immune responses. Virulence factors are encoded on mobile genetic elements (plasmids or prophage) and can be transferred among strains by horizontal gene transfer (Otto 2014).

Toxins are poisonous substances that directly harm the host. Among those secreted by *Staphylococcus aureus* there are membrane-damaging toxins, toxins that interfere with receptor functions, and enzymes that interfere with important host defense mechanisms.

Membrane-damaging toxins are cytolytic and cause pores in the cytoplasmic membrane provoking the efflux of metabolites. They can be receptor-mediated with high cell specificity, such as hemolysins and leukotoxins that lyse red blood cells and white blood cells respectively, and the alpha-toxin responsible for the lysis of red blood cells, of a few types of leukocytes and for causing apoptosis in human monocytes, T and B cells. *Staphylococcus aureus* also secretes bicomponent toxins that require a receptor interaction for their cytolytic activity such as the Panton-Valentine leukocidin, the leukocidins LukDE and LukAB (LukGH), and the gamma-toxin (gamma-hemolysin, HlgA, HlgB, HlgC).

Membrane-damaging toxins can be also non-receptor mediated with less cell specificity, such as the delta-toxin (delta-hemolysin), part of the phenol-soluble modulins (PSMs). PSMs apart from having a nonspecific cytolytic activity can also trigger inflammatory responses and contribute to neutrophil lysis after phagocytosis, which is a mechanism of relevant importance for some highly toxic *Staphylococcus aureus* strains that in this way evade innate host defenses. Delta-toxin also induces mast cell degranulation, causing atopic dermatitis (Guerra et al. 2017; Otto 2014).

Among the toxins that interfere with receptor activities there are enterotoxins that affecting intestine functions are usually the cause for emesis and diarrhea. Enterotoxins' mechanisms of action are not well understood, but they include the triggering of the T cell activation and proliferation through a nonspecific interaction of the class II major histocompatibility complex with T cell receptors, and the activation of cytokine release with the consequence of cell death by apoptosis. *Staphylococcus aureus* strains produce several enterotoxins and enterotoxin-like toxins such as the staphylococcal enterotoxin B (SEB), considered as a biological warfare weapon; the staphylococcal enterotoxin C (SEC) responsible for infective endocarditis, sepsis, and kidney injury; and the toxic shock syndrome toxin (TSST) that provokes toxic shock syndrome (TSS) by stimulating the release of IL-1, IL-2, TNF-α, and other cytokines. TSS is a potentially fatal disease; its symptoms go from fever to rash, muscle pain, nausea, decreased blood pressure, respiratory distress, and in some cases even organ failure. If cured this syndrome has a low mortality rate of 1–3%; if left untreated the mortality increases up to 70%. In 1980 TSS was associated with the use of women's tampons due to their chemical composition. Though less frequently, this disease can occur in women after giving birth, in both men and women after a surgical wound infection, and as a complication of influenza (Otto 2014; Freeman-Cook and Freeman-Cook 2006).

Staphylococcus aureus secretes a series of other proteins that help the pathogen to evade recognition and prevent subsequent activation of the immune system, such as the CHIPS (chemotaxis inhibitory protein of *Staphylococcus aureus*).

Among the numerous enzymes that interfere with host defense mechanisms *Staphylococcus aureus* secretes several nonspecific proteases. Usually proteases degrade host proteins leading to tissue destruction, as in the case of collagenase that disrupts collagen, and/or interfere with host metabolic or signaling cascade. Aureolysin cleaves several proteins, inactivates PSMs, and promotes the maturation of other nonspecific *Staphylococcus aureus* exoproteases, as the glutamyl endopeptidase SspA. These proteases, together with the cysteine proteases staphopain A and B interfering with complement factors, lead to the evasion of complement-mediated immune response. *Staphylococcus aureus* also secretes a series of serine proteases. The biological function of this category of proteases is not clear, apart from the exfoliative toxin serine proteases responsible for the staphylococcal scalded skin syndrome (SSSS), a severe skin disease characterized by rash, blisters, and lesions of the skin on the major part of the body, with the most superficial layers of the skin (especially of hands and feet) peeling off. This disease affects children more frequently than adults, but children have a lower mortality rate of about 3%, whereas the mortality rate in adults is about 50% (Otto 2014).

Staphylococcus aureus secretes staphylokinases, coagulases, and nucleases. Staphylokinase is an enzyme that facilitates the bacterial penetration through the skin barrier, focusing on the fibrin meshwork and decreasing its ability to localize a staphylococcal infection. Staphylokinase also cleaves the complement factor C3b enhancing the attack of *Staphylococcus aureus* secreted host-damaging factors such as the fibrinogen-binding protein (Efb) and the staphylococcal complement inhibitor (SCIN), which are potent inhibitors of the C3 convertase, a key enzyme in the complement pathway (Otto 2014; Plata et al. 2009; Foster 2005; Fedtke et al. 2004; Foster 2004).

Staphylocoagulase and the von Willebrand factor (vWF) are two coagulases that have the aim to form fibrin clots on the surface of the bacterial cells, thus inhibiting phagocytosis and in some cases causing the formation of abscesses and promoting the adhesion of *Staphylococcus aureus* cells to the catheter surfaces during biofilm-associated infections.

Staphylococcal nucleases seem to decrease the antibacterial activity of neutrophil extracellular traps (NETs), lipases of which little is known, and a sphingomyelinase called beta-toxin that deteriorates the sphingomyelin present on the surface of several host cells (Otto 2014).

Another factor regarding the pathogenicity of *Staphylococcus aureus* is its responsibility for food contamination and consequent intoxication. When food is left at room temperature for long periods of time *Staphylococcus aureus* cells grow and start the production of staphylococcal enterotoxins. All the staphylococcal enterotoxins are heat and protease resistant, so after the consumption of contaminated food, toxins are absorbed and bacteria pass through the intestine without causing any adverse effects on the host. Symptoms of staphylococcal intoxication appear within 1–6 h and subside within 24 h; they include nausea, emesis, abdominal pain, diarrhea, headache, cramping, and anaphylactic shock. Staphylococcal enterotoxins also damage the intestinal epithelial cells causing villus distension, crypt elongation, and lymphoid hyperplasia. Sometimes the intoxication is provoked by an aerosol exposure; in this case symptoms are sudden fever that can persist for several days, chills, headache, and cough that can last even for 2 weeks (Bhunia 2008; Freeman-Cook and Freeman-Cook 2006).

Staphylococcus aureus wide pathogenetic mechanisms are particularly enhanced by vesicles, both intracellular and extracellular, that in Gram-positive bacteria are called membrane vesicles.

Membrane Vesicles

Studies regarding the production of membrane vesicles by Gram-positive bacteria started to appear just in 1990; a major interest was registered in the last two decades, when it became clear that extracellular vesicles are a common feature of all domains of life and are present in Archaea, Bacteria (both Gram-negative and Gram-positive), and Eukaryotes. Membrane vesicles gained attention especially for their key role in bacterial pathogenesis, such as spreading of antibiotic resistance, nutrient uptake, transfer of nucleic acid, bacterial physiology, and disease progression. Recently, due to their peculiarities, interest was also raised to the prospect of using membrane vesicles as new vaccine vectors (Briaud and Carroll 2020; Wang et al. 2018).

Gram-positive bacteria produce spherical, bi-layered membrane vesicles with diameters from 20 nm to 400 nm; the membrane vesicles produced by *Staphylococcus aureus* are smaller with diameters ranging approximately from 20 nm to 130 nm (Fig. 1.2). So far it seems that every *Staphylococcus aureus* strain has the ability to produce membrane vesicles both *in vitro* and *in vivo* (Cao and Lin 2021; Tartaglia et al. 2020; Askarian et al. 2018; Curia et al. 2017; Erega et al. 2017; Brown et al. 2015; Gurung et al. 2011; Lee et al. 2009). The bacterial growth conditions (growth medium composition, temperature, availability of iron and oxygen, exposure to antibiotics, and stress) strongly influence the production of membrane vesicles, the rate of vesicle formation, their size, and the kinds of proteins associated with them (Cao and Lin 2021; Toyofuku et al. 2019; Askarian et al. 2018).

Due to their nanometer size membrane vesicles require specific imaging methods to be visualized, the most common being represented by electron microscopy, where samples usually need to be adequately prepared prior to imaging, although optical microscopy imaging techniques for real-time observation are being developed. The continuous research of good imaging methods is obviously finalized to obtain a good picture and is associated to the issue of image manipulation. In the case of microscopy images, manipulation is far from being a temptation relegated to post-production; instead it is necessary and inevitable during the acquisition process, represented by all the choices that the microscopist takes when imaging (Rossner and Yamada 2004). Imaging membrane vesicles plays a key role in the investigation of membrane vesicles' spatiotemporal properties: *in vitro* imaging helping to understand the physical characteristics and the mechanisms of release and uptake; *in vivo* imaging being useful to study their pharmacokinetic properties to utilize membrane vesicles as drug vehicles (Chuo et al. 2018; Jung and Mun 2018).

Membrane vesicles are generated in a process called vesicle biogenesis or vesiculogenesis that in the case of Gram-positive bacteria is still not completely

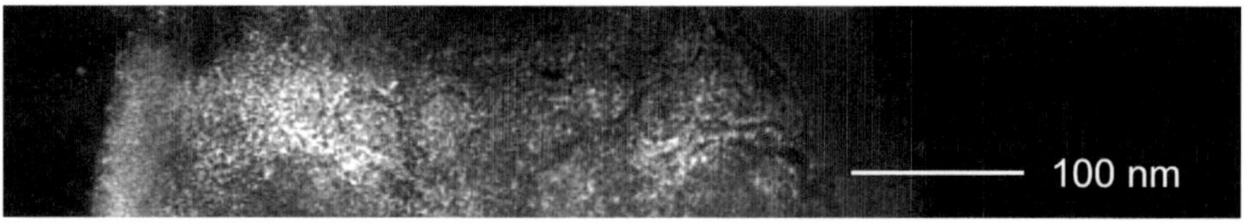

Fig. 1.2 STEM image of a portion of a *Staphylococcus aureus* cell after incubation with polyurethane. Membrane vesicles and polyurethane nanoparticles (white dots) are clearly visible. Polyurethane nanoparticles are present inside the membrane vesicles, in the cytoplasm, on the cell wall, and outside the cell

understood. The first articles regarding the discharge of membrane vesicles described the budding of spherical vesicles from the bacterial surface and a subsequent appearance of the secreted membrane vesicles in the medium (Lee 2012; Gurung et al. 2011). It is now hypothesized that the vesicle release is composed of two steps. The first one is represented by the budding of the cytoplasmic membrane at precise lipid-enriched domains; critical elements in this passage are the turgor pressure and the membrane fluidity, a key factor for an increased discharge of membrane vesicles. The effective passage through the cell wall is the final step of the shedding process. Alterations in the cell wall play a fundamental role in the vesicle release; for instance a reduced degree of peptidoglycan cross-linking results in an increment of released membrane vesicles (Cao and Lin 2021; Briaud and Carroll 2020; Tartaglia et al. 2020; Askarian et al. 2018; Brown et al. 2015).

Another route of membrane vesicle formation observed in Gram-positive bacteria is the one triggered by the action of endolysins. The production of these proteolytic enzymes is increased in bacteria infected by phages to promote the lysis of the bacterial cell wall and the consequent release of phage progeny, and to enhance the formation and discharge of membrane vesicles. Phage-triggered cell lysis is a rather new vesicle formation route that up to now seems to be still underestimated (Toyofuku et al. 2019).

During the formation process several components are enclosed into membrane vesicles: enzymes, lipids, nucleic acids, pathogenesis-associated proteins, extracellular and membrane-associated virulence factors, toxins, and immune evasion proteins. The study of the core proteome of membrane vesicles derived from *Staphylococcus aureus* pointed out that 47% of the proteins identified were common to several strains, suggesting that the packaging of proteins in membrane vesicles is not a random process, but the consequence of a conserved selective mechanism not yet completely clarified (Tartaglia et al. 2020; Brown et al. 2015; Lee et al. 2009).

More specifically membrane vesicles can contain nucleases, proteases, amidases, surface antigens and immunoglobulin-binding protein, proteins involved in metal ion acquisition, ribosomal proteins, DNA polymerases, tRNA synthetases, metabolic enzymes, lipoproteins, PSMs, hemolysins, leukocidins, elastin-binding proteins, autolysins, adhesins, proteolysins, coagulases, peptidoglycan precursors, extracellular adherence proteins, lipases, superantigens, IgG-binding proteins, and antibiotic resistance-associated proteins such as β-lactamases and penicillin-binding proteins.

Membrane vesicles, being a secretory route to deliver virulence factors to hosts, have a key role in the bacterial pathogenesis, enhancing virulence, drug resistance, host cell invasion, immune system evasion, and spreading of systemic infections in the host. The specific analysis of the cargoes is helpful to describe the potency of bacteria and to determine the biological activities of the vesicles: membrane vesicles deliver toxins to host cells provoking host inflammatory responses; membrane vesicles containing antibiotic-degrading enzymes able to inactivate the antibiotic action confer transient resistance to surrounding bacteria; membrane vesicles containing superantigens induce the activation of human T cells, whereas membrane vesicles containing immune evasion proteins are helpful to evade the immune system; membrane vesicles containing hemolysins can induce apoptosis; membrane vesicles containing coagulation factors contribute to the protection of bacteria from the host defense mechanisms; membrane vesicles containing proteins for the ion acquisition prevent the ion starvation during the host infection; other proteins are involved in nutrient transport, in the activation of the Toll-like receptor 2, and in the production of biofilm. For their ability to transport active molecules, protecting them from a hostile environment, membrane vesicles are also important for the cell-to-cell communication; the production of membrane vesicles also helps bacteria to adapt to changes in the environment and to overcome stress situations carrying stress substances out of the bacteria (Cao and Lin 2021; Briaud and Carroll 2020; Tartaglia et al. 2020; Askarian et al. 2018; Brown et al. 2015; Gurung et al. 2011; Lee et al. 2009).

Though the roles played by membrane vesicles in *Staphylococcus aureus* pathogenesis are still not completely characterized, it seems that the strategy used by *Staphylococcus aureus* for delivering bacterial components to host cells via membrane vesicles is common for both animal and human *Staphylococcus aureus* strains. Once inside the host cell, membrane vesicles play an important role in tissue invasion, adhesion, and colonization. The pathogenesis of the infection can be mediated in different ways: membrane vesicles can be cytotoxic to host cells, induce the production of cytokines, contribute to biofilm formation, mediate antibiotic resistance, or increase the bacterial survival in human blood. It seems that a rather high number of membrane vesicles delivered to the host protect *Staphylococcus aureus* bacterial cells from being killed by human blood cells as neutrophils. Moreover *Staphylococcus aureus*-derived membrane vesicles present extracellular or surface-associated virulence proteins that have a relevant role in several pathologies (Briaud and Carroll 2020; Tartaglia et al. 2020; Askarian et al. 2018; Lee 2012; Gurung et al. 2011; Lee et al. 2009).

Bacterial-derived membrane vesicles can provoke systemic infections to the host or trigger pathologies such as atopic dermatitis. According to the bacterial strain from which membrane vesicles derive, a different kind of host response is induced. Other elements that influence the host response are the number of membrane vesicles delivered (the response is dose-dependent), the type of cargo packaged during vesiculogenesis, and the integrity of membrane vesicles; in fact it has been demonstrated that lysed vesicles do not cause cytotoxicity (Briaud and Carroll 2020; Askarian et al. 2018; Gurung et al. 2011). The delivery of membrane vesicles also affects the interspecies communication; when delivered to host cells *Staphylococcus aureus* membrane vesicles can confer antibiotic resistance to other already present bacteria even though they belong to different species, providing a bacterial cooperation aimed at spreading the antibiotic resistance (Brown et al. 2015; Lee et al. 2013). An element of utter importance is that membrane vesicles are a good mechanism to deliver toxins and other molecules at high concentrations: the cargo is protected by the membrane and hence is not diluted.

The production of membrane vesicles is an advantage for bacteria that in this way are able to respond to the environment, adapting to adverse and stressful conditions, carrying stress substances out of the bacterial cell (Cao and Lin 2021).

Membrane vesicles are also a good strategy to attack the host from the inside, directly releasing the noxious cargo into the host cell. There is a multiplicity of modes for membrane vesicle delivery to host cells currently hypothesized such as dynamin-dependent endocytosis, via cholesterol-rich membrane micro-domains, and clathrin-dependent endocytosis (Alzahrani and Saadeldin 2021; Segev 2009; Honeyman et al. 2001). The size of membrane vesicles and the type of host cells attacked influence both the routes of membrane vesicles and their mode of entry; thus it is possible that unknown modes of delivery exist (Briaud and Carroll 2020). So far it has been observed that the bacterial cytoskeleton plays a major role in the cell-to-cell communication, especially in the contact-dependent signaling mechanisms used by bacteria to control their virulence; this system can be inter- or intraspecific and allows the coexistence of both the bacterium and the host (Singhi and Srivastava 2020; Mannherz 2017). Unfortunately, little is known about the role played by the bacterial cytoskeleton in the guidance of membrane vesicles, and more investigations are needed to also comprehend what happens when membrane vesicle enters the host cell and how the vesicle is received by the host cytoskeletal structures.

Another element of relevance is represented by the protein corona, the heterogeneous group of proteins and ions that in biological fluids binds to the surface of nanoparticles, driving the interactions of such nanoparticles with other particles and cells. For their nanometer size membrane vesicles are considered as nanoparticles and according to the composition of their plasma membrane can be coated by a unique protein corona. The presence of the protein coating significantly determines the interaction of nanoparticles and living systems, also interfering with mechanisms such as nanoparticle uptake. The protein corona layer grows significantly over time and can critically affect the processes of uptake and secretion of membrane vesicles for both bacteria and hosts (Wang et al. 2021; Lundqvist 2013; Tenzer et al. 2013).

Membrane vesicles have a key role also within biofilms, one the principal microbial virulence mechanisms. Membrane vesicles, particularly abundant in the biofilm matrices, confer protection against the host immune response and antibiotic treatments to microorganisms (Cao and Lin 2021; Campoccia et al. 2021; Toyofuku et al. 2019; Brown et al. 2015).

Biofilm

Staphylococcus aureus has the ability to produce biofilm, a community where bacterial cells adhere to a substratum and to each other, creating a multi-layered cellular structure embedded in an extracellular matrix (ECM) in which cells communicate via quorum sensing.

Biofilm formation starts with the adhesion, promoted by the mediation of bacterial surface proteins of bacterial cells to a polymeric surface, and proceeds with the production of the ECM and the creation of bacterial microcolonies onto the artificial material. After this step, known as maturation phase, all the bacterial cells are embedded in an array of extracellular polymeric substances (EPS). This process ends with the dispersal of the biofilm through the detachment of nomadic bacterial cells or clusters of cells that can happen when bacteria return to a planktonic lifestyle, due to sloughing or enzymatic digestion (Karygianni et al. 2020; Nguyen et al. 2020; Arciola et al. 2018; Curia et al. 2017; Costerton et al. 2005).

Due to the necessity to create the biofilm morphotype that better adapts to determined conditions, the structure and composition of biofilm vary depending on several factors: the nature of the microorganism is of primary importance, in fact even belonging to the same species; each strain produces a unique biofilm, and monospecies or multispecies biofilm are differently organized. Other determinant factors are related to the environmental conditions, to the possible stress for the cells, and to the nutrients availability.

The chemically heterogenous ECM provides mechanical stability and a biochemical microenvironment that includes exopolysaccharides, nucleic acids, proteins, lipids, membrane vesicles, and other biomolecules such as virulence factors, fundamental for the biofilm.

A key part of the EPS is represented by exopolysaccharides that contribute to bacterial pathogenicity; *Staphylococcus aureus* only produces a polysaccharide intercellular adhesin (PIA) or poly-β(1–6)-N-acetylglucosamine (PNAG), which synthesis is mediated by the *ica* operon. For its role in adhesion, cohesion, scaffolding, and stability, PIA significantly contributes to the staphylococcal biofilm formation; PIA also confers protection against antibiotics and slows down neutrophil recruitment and bacterial clearance by the host. Among the EPS dispersed in the matrix of a *Staphylococcus aureus* biofilm, there are several proteins: the fibronectin-binding proteins; the *Staphylococcus aureus* surface protein G and the biofilm-associated protein that mediate adhesion, cell-to-cell binding, scaffolding, and stability; PSMs responsible for lysing host cells and for biofilm spreading; and the staphylococcal protein A that beyond the function of adhesion has a role in immune evasion. Teichoic and lipoteichoic acids are also part of the staphylococcal EPS with the role of adhesion, cohesion, protection, and immune evasion (Campoccia et al. 2021; Karygianni et al. 2020; Nguyen et al. 2020; Graf et al. 2019; Plata et al. 2009; Costerton et al. 2005).

The EPS, which represent a nutrient reservoir for bacterial cells, are continuously produced by bacteria in biofilm, thus promoting the microbial cell-cell cohesion, the three-dimensional expansion of the matrix, and the formation of clusters of bacterial cells, also called microcolonies. EPS are of extreme importance in a biofilm because they provide biofilm integrity, physical stability, and resistance to mechanical removal; EPS components reacting with antimicrobials and limiting their access into the deeper layers of cells also confer tolerance against drugs to the biofilm. Lately it has been found out that the extracellular DNA (eDNA), part of the EPS of both environmental and clinical biofilms, not only has the ability to create supramolecular structures as stable filamentous networks through the matrix, but also mediates horizontal gene transfer, restricts the diffusion of antimicrobials, promotes antibiotic-resistant phenotypes, interacts with cells of the host immune system conditioning both the innate and the cell-mediate immune response, guides biofilm spreading, and serves as a nutrient source during starvation. Moreover eDNA interacts with factors expressed on the bacterial surfaces and with abiotic surfaces, thus promoting bacterial aggregation, mediating the bacterial attachment to abiotic surfaces, and stabilizing the architecture of the biofilm (Campoccia et al. 2021; Karygianni et al. 2020; Graf et al. 2019).

Bacterial biofilm-associated infections are estimated to be between 65% and 80% of all human bacterial infections; it is particularly critical to treat this kind of infections with conventional antibiotics because bacteria can elude innate and adaptive host immunity, and also biocides, antibiotics, antimicrobials, and chemotherapies, thus spreading antibiotic resistance. In light of this the ability of one of the most common nosocomial pathogens as *Staphylococcus aureus* to form biofilms on host tissues and medical devices represents a major healthcare problem: the bacterial attack affects the material's stability leading to the failure of the prosthesis and to the need to replace the device; moreover new surgeries increase the danger of recalcitrant and recurrent disease provoked by groups of persister cells still alive due to an insufficient antibiotic cure.

Bacterial-fungal biofilm-associated coinfections are not rare on implanted medical devices and often the cooperation between bacteria and fungi enhances the severity of the disease. In the case of *Staphylococcus aureus* and *Candida albicans* infections, the two microorganisms adopt strategies to help each other to create damages on the organ walls in order to better penetrate the host organism and to evade the phagocytic killing mediated by the opsonizing activity of human polymorphonuclear leukocytes (PMNs) (Campoccia et al. 2021; Graf et al. 2019; Arciola et al. 2018; Curia et al. 2017; Nair et al. 2014; Plata et al. 2009; Costerton et al. 2005; Fedtke et al. 2004).

Biodestruction

Biofilm infections associated to medical devices can highly affect the implant' stability and lead to the failure of the device. The categories of clinical implants mostly interested by biofilm infections vary from intravascular catheters, heart valves, coronary stents, ventricular shunts, arthro-prostheses, breast implants, cochlear implants, intra-ocular lenses, and dental implants.

The vital activity of microorganisms embedded in a biofilm inevitably deteriorates the device, causing changes in the properties of the material and leading to its biodestrucion (Arciola et al. 2018; Curia et al. 2017; Didenko et al. 2012; Costerton et al. 2005).

The starting point of this book is the focus on the biodestructive action of *Staphylococcus aureus* on polyurethane prostheses used in orthopedic stomatology. It is notable to remind that *Staphylococcus aureus* is a common component of the oral flora, often responsible for chronic inflammations of the soft tissues of the oral cavity such as periodontitis and gingivitis, able to influence the development of caries. The high pathogenicity of *Staphylococcus aureus* and its ability to cause chronic infectious processes and to form biofilm lead to the deterioration of polyurethane dental prostheses (Curia et al. 2017; McCormack et al. 2015; Didenko et al. 2012).

Polyurethane is characterized by synthetic versatility, excellent mechanical properties (endurance, strength, and abrasion resistance), and good biocompatibility; these features make polyurethane one of the most versatile plastic materials that can be adapted to meet specific needs. The fields of application of polyurethane are various and space from coatings to paints, sealants, construction and insulation materials, kitchen appliances, adhesives, fibers, packaging, and clinical applications as general-purpose tubing, wound dressings, and drug delivery.

Structurally polyurethane is a polymer formed by the condensation of polyol and polyisocyanate, with the repeating unit containing a urethane linkage. Polyurethanes with different characteristics can be synthesized varying the type of polyisocyanate and polyol; changing the number of substitutions and the spacing between and within branch chains, it is possible to produce linear or branched polyurethanes with different degrees of tensile strength and elasticity (Cooper and Guan 2016; Shah et al. 2008; Howard 2002; Polyurethane n.d. www.americanchemistry.com/chemistry-in-america/chemistries/polyurethane).

Due to their chemical properties, surface charge, hydrophobicity, and surface roughness, polyurethane medical implants are susceptible to the microbial adhesion and consequent biofilm formation. The microorganisms attack the polymer hydrolyzing the ester bond, operating a polyurethanolytic action that results in the biodegradation of the material. Two types of polyurethane-esterases have been characterized: a cell-associated membrane-bound polyurethane-esterase that provides access to the hydrophobic polyurethane surface, and an extracellular polyurethane-esterase that then attaches to the polyurethane exposed surface. Thanks to the combined action of the two enzymes, microorganisms manage to adhere to the polyurethane surface and hydrolyze the substrate to obtain metabolites (Bhavsar et al. 2023; Shah et al. 2008).

The microbial activity of *Staphylococcus aureus*, particularly enhanced by the biofilm condition, adversely affects the strength and durability of polyurethane provoking damages and causing structural changes on the prostheses. Eventually the bacterial action on the urethane bond results in the biodestruction of the dental implant with consequent detachment of micro- and nano-debris from the bulk material (Fig. 1.3). The biodestruction operated by *Staphylococcus aureus* generates a new variety of nanoparticle, very different from the engineered polyurethane nanoparticles commonly used in several industrial and medical applications, which properties need to be studied. The newly generated nanoparticles remain embedded in the biofilm ECM so that during the dispersal of the biofilm polyurethane nanoparticles are dragged away with the detaching bacterial cells (Fig. 1.4).

This process together with the consequences that the biodestruction of implants could cause on the human body is not yet completely understood. Questions about the human health, the destiny of the polyurethane nanoparticles within the body, the interactions with other organs and tissues, and the possible toxicity that nanoparticles derived from a xenobiotic material have, are still unanswered, so investigating the whole microbial biodestructive process remains of paramount importance.

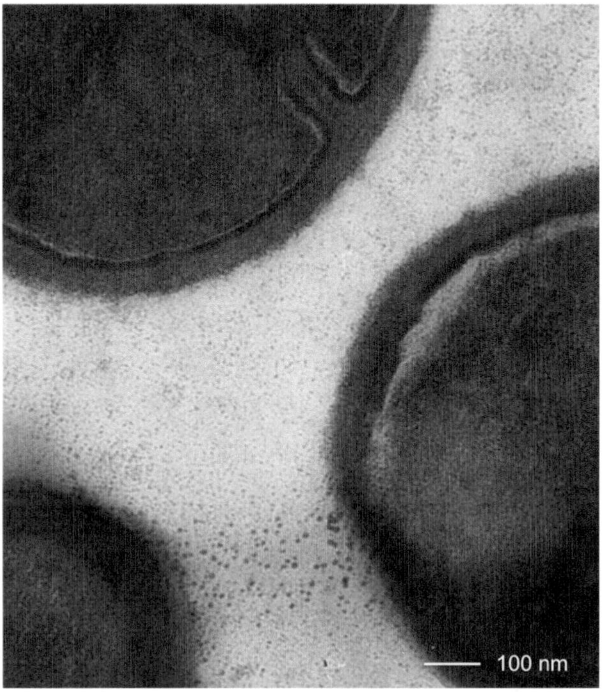

Fig. 1.3 TEM image of portions of three *Staphylococcus aureus* cells after incubation with polyurethane. The sample was prepared by lead citrate and uranyl acetate was used for contrasting. Polyurethane nanoparticles (black dots) are present in the medium; in restricted areas associated to the minimum distance between couples of bacterial cells, polyurethane nanoparticles show a higher density, suggesting the presence of nanoparticles trafficking between neighbor cells

Fig. 1.4 SEM image of clusters of *Staphylococcus aureus* cells incubated on a polyurethane surface after a 7-day incubation. Polyurethane nanoparticles (white dots) are regularly distributed on the bacterial surfaces

References

Alzahrani FA, Saadeldin IM. Role of exosomes in biological communication systems. Singapore: Springer; 2021. https://doi.org/10.1007/978-981-15-6599-1.

Arciola CR, Campoccia D, Montanaro L. Implant infections: adhesion, biofilm formation and immune evasion. Nat Rev Microbiol. 2018;16:397–409. https://doi.org/10.1038/s41579-018-0019-y.

Askarian F, Lapek JD Jr, Dongre M, Tsai CM, Kumaraswamy M, Kousha A, Valderrama JA, Ludviksen JA, Cavanagh JP, Uchiyama S, Mollnes TE, Gonzalez DJ, Wai SN, Nizet V, Johannessen M. *Staphylococcus aureus* membrane-derived vesicles promote bacterial virulence and confer protective immunity in murine infection models. Front Microbiol. 2018;9:262. https://doi.org/10.3389/fmicb.2018.00262.

Bhavsar P, Bhave M, Webb HK. Solving the plastic dilemma: the fungal and bacterial biodegradability of polyurethanes. World J Microbiol Biotechnol 2023;39:122. https://doi.org/10.1007/s11274-023-03558-8.

Bhunia AK. Staphylococcus aureus. In: Bhunia AK, editor. Foodborne microbial pathogens. New York: Springer; 2008. p. 125–34. https://doi.org/10.1007/978-0-387-74537-4_6.

Briaud P, Carroll RK. Extracellular vesicle biogenesis and functions in gram-positive bacteria. Infect Immun. 2020;88(12):e00433–20. https://doi.org/10.1128/IAI.00433-20.

Brown L, Wolf JM, Prados-Rosales R, Casadevall A. Through the wall: extracellular vesicles in gram-positive bacteria, mycobacteria and fungi. Nat Rev Microbiol. 2015;13(10):620–30. https://doi.org/10.1038/nrmicro3480.

Campoccia D, Montanaro L, Arciola CR. Extracellular DNA (eDNA). A major ubiquitous element of the bacterial biofilm architecture. Int J Mol Sci. 2021;22(16):9100. https://doi.org/10.3390/ijms22169100.

Cao Y, Lin H. Characterization and function of membrane vesicles in gram-positive bacteria. Appl Microbiol Biotechnol. 2021;105:1795–801. https://doi.org/10.1007/s00253-021-11140-1.

CDC. Etymologia: *Streptococcus*. Emerg Infect Dis. 2016;22(11):1977. https://doi.org/10.3201/eid2211.et2211.

Cheng A, DeDent AC, Schneewind O, Missiakas D. A play in four acts: *Staphylococcus aureus* abscess formation. Trends Microbiol. 2011;19(5):225–32. https://doi.org/10.1016/j.tim.2011.01.007.

Chuo STY, Chien JCY, Lai CPK. Imaging extracellular vesicles: current and emerging methods. J Biomed Sci. 2018;25:91. https://doi.org/10.1186/s12929-018-0494-5.

Cooper SL, Guan J, editors. Advances in polyurethane biomaterials. Duxford: Woodhead Publishing; 2016.

Costerton JW, Montanaro L, Arciola CR. Biofilm in implant infections: its production and regulation. Int J Artif Organs. 2005;28(11):1062–8. https://doi.org/10.1177/039139880502801103.

Curia R, Milani M, Didenko LV, Avtandilov GA, Shevlyagina NV, Tatti F. Analysis of the bacterial vesicles' enhanced toxicological threat via electron microscopy. In: Talukdar K, Bhushan M, Malipatil AS, editors. Advancing medicine through nanotechnology and nanomechanics applications. Hershey: IGI Global; 2017. p. 19–43. https://doi.org/10.4018/978-1-5225-1043-7.ch002.

Didenko LV, Avtandilov GA, Shevlyagina NV, Smirnova TA, Lebedenko IY, Tatti F, Savoia C, Evans G, Milani M. Biodestruction of polyurethane by *Staphylococcus aureus* (an investigation by SEM, TEM and FIB). In: Méndez-Vilas A, editor. Current microscopy contributions to advances in science and technology. Badajoz: Formatex Research Center Publishing; 2012. p. 323–34.

Erega A, Curia R, Milani M. Interactions between nanoparticles and bacterial structures as seen through software enhanced analysis of electron microscopy images. In: Méndez-Vilas A, editor. Microscopy and imaging science: practical approaches to applied research and education. Badajoz: Formatex Research Center Publishing; 2017. p. 587–94.

Fedtke I, Götz F, Peschel A. Bacterial evasion of innate host defenses – the *Staphylococcus aureus* lesson. Int J Med Microbiol. 2004;294:189–94. https://doi.org/10.1016/j.ijmm.2004.06.016.

Foster T. *Staphylococcus*. In: Baron S, editor. Medical microbiology. 4th ed. Galveston: University of Texas Medical Branch at Galveston; 1996, Chapter 12.

Foster TJ. The *Staphylococcus aureus* "superbug". J Clin Invest. 2004;114(12):1693–6.

Foster TJ. Immune evasion by *Staphylococci*. Nat Rev Microbiol. 2005;3(12):948–58. https://doi.org/10.1038/nrmicro1289.

Fraunholz M, Sinha B. Intracellular *Staphylococcus aureus*: live-in and let die. Front Cell Infect Microbiol. 2012;2:43. https://doi.org/10.3389/fcimb.2012.00043.

Freeman-Cook L, Freeman-Cook KD. *Staphylococcus aureus* infections. New York: Infobase Publishing; 2006.

Graf AC, Leonard A, Schäuble M, Rieckmann LM, Hoyer J, Maass S, Lalk M, Becher D, Pané-Farré J, Riedel K. Virulence factors produced by *Staphylococcus aureus* biofilms have a moonlighting function contributing to biofilm integrity. Mol Cell Proteomics. 2019;18:1036–53. https://doi.org/10.1074/mcp.RA118.001120.

GS. Great teachers of surgery in the past - Alexander Ogston (1844–1929). Br J Surg. 1965;52:917–20. https://doi.org/10.1002/bjs.1800521203.

Guerra FE, Borgogna TR, Patel DM, Sward EW, Voyich JM. Epic immune battles of history: neutrophils vs. *Staphylococcus aureus*. Front Cell Infect Microbiol. 2017;7:286. https://doi.org/10.3389/fcimb.2017.00286.

Gurung M, Moon DC, Choi CW, Lee JH, Bae YC, Kim J, Lee YC, Seol SY, Cho DT, Kim SI, Lee JC. *Staphylococcus aureus* produces membrane-derived vesicles that induce host cell death. PLoS One. 2011;6(11):e27958. https://doi.org/10.1371/journal.pone.0027958.

Harris LG, Foster SJ, Richards RG. An introduction to *Staphylococcus aureus*, and techniques for identifying and quantifying *Staphylococcus aureus* adhesins in relation to adhesion to biomaterials: review. Eur Cells Mater. 2002;4:39–60. https://doi.org/10.22203/eCM.v004a04.

Hassan AY, Lin JT, Ricker N, Anany H. The age of phage: friend or foe in the new dawn of therapeutic and biocontrol applications? Pharmaceuticals. 2021;14:199. https://doi.org/10.3390/ph14030199.

Honeyman A, Friedman H, Bendinelli M. *Staphylococcus aureus* infection and disease. New York: Springer Science+Business Media; 2001. https://doi.org/10.1007/b111097.

Howard GT. Biodegradation of polyurethane: a review. Int Biodeterior Biodegradation. 2002;49:245–52. https://doi.org/10.1016/S0964-8305(02)00051-3.

Hrebík D, Štveráková D, Škubník K, Füzik T, Pantůček R, Plevka P. Structure and genome ejection mechanism of *Staphylococcus aureus* phage P68. Sci Adv. 2019;5:eaaw7414.

Jung MK, Mun JY. Sample preparation and imaging of exosomes by transmission electron microscopy. J Vis Exp. 2018;131:e56482. https://doi.org/10.3791/56482.

Karygianni L, Ren Z, Koo H, Thurnheer T. Biofilm matrixome: extracellular components in structured microbial communities. Trends Microbiol. 2020;28(8):668–81. https://doi.org/10.1016/j.tim.2020.03.016.

Lee EY, Choi DY, Kim DK, Kim JW, Park JO, Kim S, Kim SH, Desiderio DM, Kim YK, Kim KP, Gho YS. Gram-positive bacteria produce membrane vesicles: proteomics-based characterization of *Staphylococcus aureus*-derived membrane vesicles. Proteomics. 2009;9(24):5425–36. https://doi.org/10.1002/pmic.200900338.

Lee J, Lee EY, Kim SH, Kim DK, Park KS, Kim KP, Kim YK, Roh TY, Gho YS. *Staphylococcus aureus* extracellular vesicles carry biologically active β-lactamase. Antimicrob Agents Chemother. 2013;57(6):2589–95. https://doi.org/10.1128/AAC.00522-12.

Lee JC. *Staphylococcus aureus* membrane vesicles and its potential role in bacterial pathogenesis. J Bacteriol Virol. 2012;42(3):181–8. https://doi.org/10.4167/jbv.2012.42.3.181.

Licitra G. Etymologia: *Staphylococcus*. Emerg Infect Dis. 2013;19(9):1553. https://doi.org/10.3201/eid1909.et1909.

Lowy FD. Antimicrobial resistance: the example of *Staphylococcus aureus*. J Clin Invest. 2003;111(9):1265–73. https://doi.org/10.1172/JCI18535.

Lundqvist M. Tracking protein corona over time. Nat Nanotechnol. 2013;8:701–2. https://doi.org/10.1038/nnano.2013.196.

Mannherz HG. The actin cytoskeleton and bacterial infection. Cham: Springer International Publishing; 2017. https://doi.org/10.1007/978-3-319-50047-8.

McCormack MG, Smith AJ, Akram AN, Jackson M, Robertson D, Edwards G. *Staphylococcus aureus* and the oral cavity: an overlooked source of carriage and infection? Am J Infect Control. 2015;43:35–7. https://doi.org/10.1016/j.ajic.2014.09.015.

Myelnikov D. An alternative cure: the adoption and survival of bacteriophage therapy in the USSR, 1922–1955. J Hist Med Allied Sci. 2018;73(4):385–411. https://doi.org/10.1093/jhmas/jry024.

Nair N, Biswas R, Götz F, Biswas L. Impact of *Staphylococcus aureus* on pathogenesis in polymicrobial infections. Infect Immun. 2014;82(6):2162–9. https://doi.org/10.1128/IAI.00059-14.

Nguyen HTT, Nguyen TH, Otto M. The staphylococcal exopolysaccharide PIA – biosynthesis and role in biofilm formation, colonization, and infection. Comput Struct Biotechnol J. 2020;18:3324–34. https://doi.org/10.1016/j.csbj.2020.10.027.

Ogston A, Witte W. On abscesses. Rev Infect Dis. 1984;6(1):122–8. https://www.jstor.org/stable/4453267

Otto M. *Staphylococcus aureus* toxins. Curr Opin Microbiol. 2014;17:32–7. https://doi.org/10.1016/j.mib.2013.11.004.

Plata K, Rosato AE, Węgrzyn G. *Staphylococcus aureus* as an infectious agent: overview of biochemistry and molecular genetics of its pathogenicity. Acta Biochim Pol. 2009;56(4):597–612. https://doi.org/10.18388/abp.2009_2491.

Polyurethane. www.americanchemistry.com/chemistry-in-america/chemistries/polyurethane.

Rasigade JP, Vandenesch F. *Staphylococcus aureus*: a pathogen with still unresolved issues. Infect Genet Evol. 2014;21:510–4. https://doi.org/10.1016/j.meegid.2013.08.018.

Rossi G, Cerquetella M, Attili AR. Amphixenosic aspects of *Staphylococcus aureus* infection in man and animals. In: Bagnoli F, Rappuoli R, Grandi G, editors. *Staphylococcus aureus*. Current topics in microbiology and immunology, vol. 409. Cham: Springer; 2016. p. 297–323. https://doi.org/10.1007/82_2016_2.

Rossner M, Yamada KM. What's in a picture? The temptation of image manipulation. J Cell Biol. 2004;166(1):11–5. https://doi.org/10.1083/jcb.200406019.

Rowe HM, Meliopoulos VA, Iverson A, Bomme P, Schultz-Cherry S, Rosch JW. Direct interactions with influenza promote bacterial adherence during respiratory infections. Nat Microbiol. 2019;4(8):1328–36. https://doi.org/10.1038/s41564-019-0447-0.

Segev N. Trafficking inside cells: pathways, mechanisms and regulation. New York: Springer-Verlag; 2009. https://doi.org/10.1007/978-0-387-93877-6.

Shah AA, Hasan F, Hameed A, Ahmed S. Biological degradation of plastics: a comprehensive review. Biotechnol Adv. 2008;26:246–65. https://doi.org/10.1016/j.biotechadv.2007.12.005.

Singhi D, Srivastava P. Role of bacterial cytoskeleton and other apparatuses in cell communication. Front Mol Biosci. 2020;7:158. https://doi.org/10.3389/fmolb.2020.00158.

Tartaglia NR, Nicolas A, de Rezende RV, Da Luz BSR, Briard-Bion V, Krupova Z, Thierry A, Coste F, Burel A, Martin P, Jardin J, Azevedo V, Le Loir Y, Guédon E. Extracellular vesicles produced by human and animal *Staphylococcus aureus* strains share a highly conserved core proteome. Sci Rep. 2020;10:8467. https://doi.org/10.1038/s41598-020-64952-y.

Tenzer S, Docter D, Kuharev J, Musyanovych A, Fetz V, Hecht R, Schlenk F, Fischer D, Kiouptsi K, Reinhardt C, Landfester K, Schild H, Maskos M, Knauer SK, Stauber RH. Rapid formation of plasma protein corona critically affects nanoparticle pathophysiology. Nat Nanotechnol. 2013;8(10):772–81. https://doi.org/10.1038/nnano.2013.181.

Toyofuku M, Nomura N, Eberl L. Types and origins of bacterial membrane vesicles. Nat Rev Microbiol. 2019;17(1):13–24. https://doi.org/10.1038/s41579-018-0112-2.

Tzagoloff H, Novick R. Geometry of cell division in *Staphylococcus aureus*. J Bacteriol. 1977;129(1):343–50.

Wang C, Chen B, He M, Hu B. Composition of intracellular protein corona around nanoparticles during internalization. ACS Nano. 2021;15:3108–22. https://doi.org/10.1021/acsnano.0c09649.

Wang X, Thompson CD, Weidenmaier C, Lee JC. Release of *Staphylococcus aureus* extracellular vesicles and their application as a vaccine platform. Nat Commun. 2018;9:1379. https://doi.org/10.1038/s41467-018-03847-z.

Electron Microscopy

It is only after you have come to know the surface of things ... that you can venture to seek what is underneath. But the surface of things is inexhaustible.
Italo Calvino

History

The observations made by Robert Hooke and Antonie van Leeuwenhoek with the first optical microscope awakened man's will to investigate the nature of things, observing objects more and more in detail. The word microscope (from Greek μικρόν small and σκοπεῖν to look at) points out the scientists' interest in looking at the particulars of objects, focusing on the disposition and the shapes of their components. The will and need to investigate the nature of things have roots in Democritus' atomist doctrine, which stated that reality was made of atoms and void, and in the René Descartes' theory of the material universe as a *plenum*. The continuous evolution of the thought led to the subsequent denial of Democritus' vision of atoms as indivisible entities in favor of Gottfried W. von Leibniz's mathematical theory of the infinite divisibility of matter (Karlsson Rosenthal 2009; Levey 1998; Meinel 1988).

In the years between 1897 and 1932 the thirst for knowledge brought to the discovery of the three components of the atom: the electron by Joseph J. Thomson, the proton by Ernest Rutherford, and the neutron by James Chadwick. The first

M. Milani et al., *Bacterial Degradation of Organic and Inorganic Materials*,
https://doi.org/10.1007/978-3-031-26949-3_2

subatomic particles to be discovered were the electrons, initially called "corpuscles." Electrons' properties were widely studied: in 1927 George P. Thomson, son of J. J. Thomson, observed electron diffraction, confirming the wave-like properties of the electrons and corroborating Louis-Victor P. R. de Broglie wave/particle hypothesis (1924), according to which electrons acted not only as particles but as waves too, so that electrons could be accelerated in the vacuum to a speed comparable to that of light. In 1927 Hans W. H. Busch, applying the Maxwell laws (1864), discovered that electron beams' paths could be oriented with electric and magnetic fields analogously to the way light is refracted by optical lenses, *de facto* laying the theoretical basis for the electron microscope (Hawkes 2004; Artsimovitch and Loukianov 1975; Chadwick 1932; Thomson and Reid 1927; De Broglie 1924; Rutherford 1920; Thomson 1897).

In 1931 while doing experiments to verify Busch's lens theory Ernst A. F. Ruska and Max Knoll built the first transmission electron microscope (TEM). By the end of the 1930s TEMs reached a resolution limit of 10 nm and in 1939 the first commercial TEM was built by Siemens. The 1940s and 1950s were decades of great improvement both for the microscope's components such as electromagnetic lenses, vacuum systems, and electron guns and for the instrumentation useful for the samples' preparation such as microtomes which provided thinner and thinner slices, allowing better results. Today TEMs have a resolution better than 0.05 nm (O'Keefe 2008; Hawkes 2004; Ruska 1987).

In 1935 Knoll described the principle of the scanning electron microscopy, and in 1942 Vladimir K. Zworykin, James Hillier, and Richard L. Snyder built a scanning electron microscope (SEM) with a resolution limit of 50 nm. The first commercial SEM was built in 1965 by Cambridge Instrument; nowadays SEMs' resolution is better than 1 nm (Ul-Hamid 2018; Rohde 2011).

In 1938 Manfred von Ardenne described the first scanning transmission electron microscope (STEM) which combined the principles of scanning and transmission microscopies. In 1969 the resolving power of the STEM was 25 nm; nowadays TEMs equipped with a STEM unit have a resolution limit of 0.05 nm in STEM mode (Liu 2021).

A significant advancement in the perspective of electron microscopy was delineated in Richard P. Feynman's speech of 1959. A lecture of great inspiration for the scientists at IBM Corporation that in the 1980s, with the aim of imaging and manipulating the matter at the nanoscale, built the scanning tunneling microscope (STM) and the atomic force microscope (AFM) (Baird and Shew 2004; Feynman 1960). Unfortunately Feynman's enthusiastic approach overlooked the complexity of the properties of nano-objects underestimating the fact that a recourse to quantum mechanics was necessary to overcome the classical approach, but anyway non sufficient.

Innovative instrumentation and new techniques viewed their development in the years from the 1960s to the 1990s: ultrahigh voltage TEMs (>300 kV) provided the electrons with a higher energy to penetrate deeper into "thick" samples; new detectors for electron energy-loss spectroscopy or X-ray spectroscopy, crucial for elemental and chemical analyses, were incorporated in SEMs turning these microscopes into more powerful instruments; new brighter sources such as the Lanthanum hexaboride filament and the field emission gun (FEG) brought better imaging and resolution; finally the possibility of tilting specimens improved examination outputs. In the early 1990s the development of the environmental scanning electron microscope (ESEM) allowed to analyze samples under natural humidity conditions, expanding to wet samples the categories of specimen that could be investigated (Stokes 2008; Bogner et al. 2007).

In 1959 Feynman not only laid the bases for nanotechnology, but in his visionary discussion he talked about the use of ion beams for sample miniaturization, actually suggesting a new type of microscope, the focused ion beam. In the early 1960s Garry Stewart fitted an ion column in a SEM chamber and Alec N. Broers improved the optics of the ion column, but it was only in the middle of the 1990s that two-beam microscopes were produced. Focused ion beam/scanning electron microscope (FIB/SEM) finally allowed the operator to observe samples using either electrons or ions, thanks to the microscope's imaging and micromachining capabilities to operate microstructural analysis and nanomachining on the specimens (Wells et al. 2012; Volkert and Minor 2007).

The microscope is man's noblest, supreme, and most far-reaching tool.
–Adrianus Pijper

The continuous development and improvement of electron microscopes aims to detailed, better quality, and more resolved images. The applications of electron microscopy are numerous and span from the investigation of cells, tissues, and microorganisms such as bacteria and viruses in life and medical sciences, to the analysis of metals, crystals, nanotubes, nanofibers, and superconductors in material sciences with industrial and commercial implementations, to soil and rock sampling in geological sciences, with applications also in the forensic field.

The principles at the basis of electron microscopy appear similar to those of the more ancient optical microscopy; in both cases the sources emit particles (massless photons or massive electrons), so the light in optical microscopes plays the role of the electron beam in electron ones, as well as glass lenses are replaced with electromagnetic ones. Photons and electrons share the wave behavior; the different wavelengths (visible light ranging from 400 nm to 700 nm; electron beam wavelength ranging from 3.88 pm at 100 keV to 2.24 pm at 300 keV) together with the different structure and components make the information acquired through optical and electron microscopes substantially diverse, with electron microscopy providing images with a richer variety of details enabling a resolution at the sub-nanometer range.

In electron microscopes electrons emitted from a gun are accelerated in the vacuum, converged into a beam by electromagnetic lenses, and either projected through a thin sample in the case of transmission microscopy or projected and swept (raster scanning) on the surface of a bulky specimen in the case of scanning microscopy. The transmitted or scattered electrons are then collected by different kinds of detectors and a digital image is acquired. The position of the electron microscopes' components, their connections, and their way of working, along with the preparation method used for the sample, are of primary importance to sense what to expect from an image and how to interpret its content.

The main components of an electron microscope are: an electron gun, electromagnetic lenses, a vacuum system, specimen chamber and holder, different types of detectors, a CCD/C-MOS (charge coupled device/complementary metal oxide semiconductor) camera, power supply system, and a computer to control all microscope's parameters and to handle and transform the acquired data into images (Egerton 2016; Candia Carnevali and Milani 2010).

Transmission Electron Microscope

In a TEM the electron beam travels through a very thin sample forming an image at the bottom of the column that is visualized on a fluorescent screen or recorded with a CCD or C-MOS camera.

A TEM consists of an electron column, a vacuum system, a high voltage generator, and a power supply system. The column is composed of an electron gun which emits the primary electron beam, an illumination system that controls the intensity and the convergence angle of the electron beam, a specimen area, an imaging system (the objective lens), and a projection system that controls the magnification of the image or diffraction pattern generated by the specimen/beam interaction. Finally, the information is visualized and recorded in the viewing chamber.

TEMs are equipped with round apertures: Molybdenum, Platinum, or Gold plates with bores ranging from 10 μm to 300 μm. There are several apertures in specific points of the beam path, each with a different task. The condenser movable aperture controls the fraction of the beam directed to the specimen, thus helping in the control of the illumination intensity. The condenser lens, together with the condenser aperture, controls the convergence angle of the beam impinging on the specimen. The objective movable aperture can be used to enhance image contrast or to allow only specific diffracted beams to form a dark-field image. Best performances require an optimal axial alignment of the filament, the illumination system lenses, and the imaging system lenses.

A TEM image is the projection of the sample's section on the fluorescent screen or camera. The contrast seen in the electron images can be described by three different modalities: mass thickness contrast, diffraction contrast, and phase contrast.

Mass Thickness Contrast

When the primary electron beam hits a thin sample a small fraction of the primary electrons is scattered due to the interaction with the atoms in the specimen. The scattering angle is directly proportional to the mass of the atomic species and depends on the samples' thickness. By centering a small aperture around the center of the diffraction pattern, the electrons scattered at high angles are blocked and the resulting image is formed only by the electrons that do not undergo scattering and pass through the aperture. The image acquired by scattering contrast is a bright-field image in which the thicker, denser, and heavier parts of the sample appear darker.

Diffraction Contrast

The interaction of the primary electron beam with periodically arranged structures generates a diffraction pattern. Electrons scattered by the sample interfere generating diffraction spots (maxima); their position and intensity depend on the lattice parameters of the crystal and on atomic species/occupancies, respectively. The selection of the diffraction spots with the objective aperture allows the formation of a dark-field image.

Phase Contrast

In a TEM a phase contrast can be acquired when a specimen is irradiated with a parallel electron beam. The contrast or phase information arises from interference between an unscattered incident wave, a zeroth-order diffraction, and scattered incident waves, representing higher-order diffractions. A phase contrast created with electrons scattered according to the Bragg effect creates a lattice image also in the complex form of moiré pattern (Mansfield et al. 2017; Egerton 2016; Ayache et al. 2010; Williams and Carter 1996; Bassett et al. 1958).

Scanning Electron Microscope

A SEM is used to investigate the topography and the morphology of a specimen' surface at a different range of magnification; acquired images give stereoscopic information thanks to the field depth. In a SEM a primary electron beam, generated by an electron gun, is focused on a small spot and scanned on the sample surface. The size of the scanned area varies with the magnification: higher the magnification, smaller the area. The interaction of the primary beam with the sample surface generates a variety of emitted signals (Fig. 2.1) that can be collected by dedicated detectors to build images of different type. The electron beam travels in vacuum towards the sample in the vacuum chamber. The primary electron beam is converged into a fine beam by electromagnetic lenses (condenser and objective lens) and, by a system of deflection coils, is deflected to raster the sample surface. The specimen chamber is equipped with a stage holder that moves the sample, and a set of detectors which collect the signals produced by the specimen.

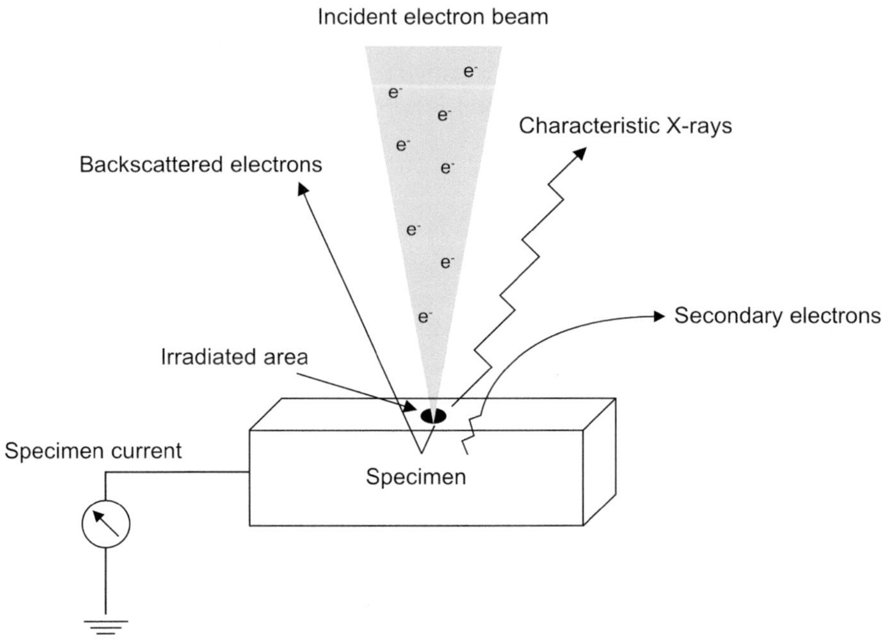

Fig. 2.1 When the sample is hit by the primary electrons (incident beam), SE and BSE are generated; they concur in the image formation together with X-ray signals for elemental analysis

When primary electrons hit the sample, they interact with the atoms composing the specimen giving rise to families of electrons: some of the incident primary electrons fly backwards at a small angle and are called backscattered electrons (BSE); other electrons travel into the sample colliding with its atoms, progressively losing energy, and generating either secondary electrons (SE) with different energies or X-rays, and at the end they remain trapped into the sample. Depending on the collected signal the image gives a specific kind of information (topographical, densitometric, or elemental).

Secondary Electrons

A nearly free electron of the specimen that undergoes inelastic collisions with primary beam's electrons or scattered ones receives part of their energy; consequently it leaves the sample and it is called a secondary electron. Only SE in the superficial layer (few tens of nanometers thick) of the sample can leave it, moving in the vacuum chamber bringing topographical information. SE and BSE are simultaneously emitted from the sample.

SE are usually detected by the Everhart and Thornley detector (ETD) invented in 1960.

Backscattered Electrons

Primary electrons elastically colliding with the atoms in the sample are scattered backwards into the vacuum chamber; they have an energy similar to the energy of primary beam electrons and are called BSE or sometimes reflected electrons. The yield of BSE emission is proportional to the atomic number of elements in the sample, so that the contrast in a BSE image is informative about the compositional distribution of the sample itself. BSE detectors can be of various types: YAG (yttrium aluminum garnet) detector, Robinson detector, semiconductor detector, or conversion signal mechanism detector.

X-Rays

When a primary electron hits an atom of the sample, ionization processes can occur with the expulsion of an electron that leaves a vacant shell. This vacancy is promptly filled by an electron from an outer orbit to re-establish a stable state. The energy conservation of this transition requires the emission of an X-ray whose energy or wavelength is unique for each element, thus providing a tool for the elemental analysis of the sample.

X-ray signal can be detected using an energy-dispersive X-ray spectrometer (EDX or EDS) or a wavelength-dispersive X-ray spectrometer (WDX or WDS) (Egerton 2016; Krumeich 2011; Ayache et al. 2010; Stokes 2008).

Scanning Transmission Electron Microscope

The principles of transmission and scanning electron microscopies are combined in the STEM (Fig. 2.2), where a convergent electron beam, scanned over the sample surface, irradiates the specimen. The detectors, placed under the sample in the plane opposite to the source, collect both the transmitted and the scattered electrons: the former give information about the density and the crystalline state of the sample, the latter about its compositional distribution. This technique can be performed in TEMs with added electron beam scanning function and STEM detectors, and in SEMs with a STEM detector mounted into the specimen chamber.

In STEMs when the electron beam hits the sample, both the transmitted and the elastically and inelastically scattered electrons follow the incident beam direction. Selecting only the transmitted electrons, blocking the scattered ones through the bright-field aperture, it is possible to acquire a bright-field STEM image where low-density areas result bright and denser parts dark. The scattered electrons, according to their angle, are collected by a dark-field detector or a high-angle annular dark-field one, which collects electrons scattered at angles higher than 50 mrad, in order to form dark-field STEM image and high-angle annular dark-field STEM image, respectively (Ayache et al. 2010).

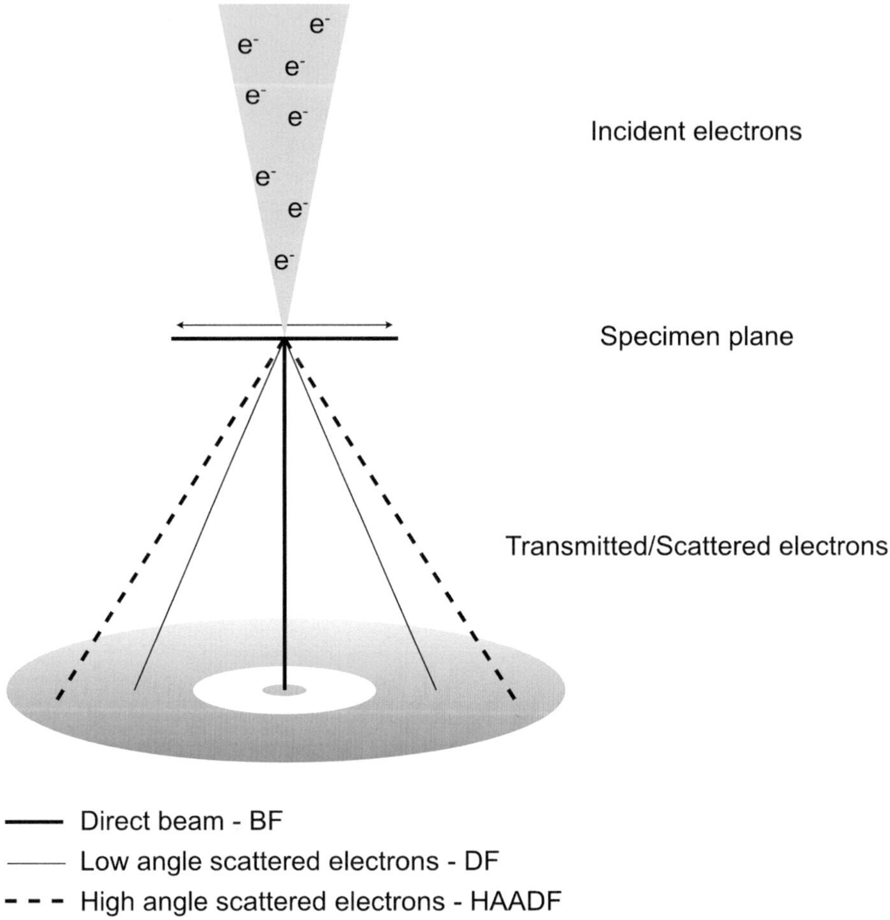

Fig. 2.2 Electrons of the incident beam after hitting the sample are scattered in different directions generating signals that, according to the angle, are collected by different detectors (bright field = BF, dark field = DF, high-angle annular dark field = HAADF) to build up images of various types of the sample

Focused Ion Beam/Scanning Electron Microscope

A FIB/SEM is a system equipped with two columns mounted onto the same specimen chamber shooting different charged particles. The SEM column is vertical as usual, and the ion column is mounted at 52°–54°. The electron source can be a tungsten filament or a FEG type; the ion source is usually emitting Gallium ions. Nowadays FIB columns, ionizing gases of different species, can also emit ions by plasma which main feature is to provide a higher beam current in the range of few μA. SEM and FIB columns can be used for imaging; thus it is possible to obtain both an ion and an electron image of the same sample' spot with different complementary information; in particular since the ion column is tilted with respect to the electron one, when imaging circular objects with the SEM, the structures result ellipsoidal in ion images. This effect has been discussed by Euclid in his theorem XXXVI: a wheel is seen sometimes as a circle, sometimes as an ellipse. The possibility to tilt the specimen can further underline this effect which must be taken into account during the interpretation of the acquired images, especially when the shape of the object is relevant for the investigation of the dynamics of the object itself (as in the case of nanoparticles).

Since ions have a larger mass than electrons, when the ion beam hits the sample, atoms and atomic clusters are sputtered from the specimen surface leaving the sample damaged and irreversibly modified not only by etching processes but also by redeposition. This operation, known as milling, creates both secondary ions, that can be also used for either imaging (ion image) or for compositional analysis, and SE used for acquiring electron microscopy images. Ion milling has a nanometric precision and removes material in a controlled way.

The polishing technique is a similar procedure in which the ion beam is scanned along the sample at lower currents than those typically used in milling, so that only the superficial layers are removed and the sample superficial flaws are eliminated.

The use of specific gases, in combination with the ion beam, accelerates etching and milling operations, whereas gases containing specific precursors allow the deposition of certain materials, creating structures on top of the surface.

A FIB/SEM system provides the possibility of sample micro-/nano-machining (production of minute components), of preparing TEM samples, and of performing a real three-dimensional reconstruction of a given volume of the sample (therefore solving image ambiguity in SEM or TEM imaging) (Milani et al. 2006; Giannuzzi 2004; Orloff et al. 2003).

Image Abnormalities

During the acquisition of electron microscopy images it is common to run up against some blurred or unfocused images and unstable brightness. Probable causes to these unwanted effects are: charge-up phenomenon responsible for mirror-like effects, contamination of the sample, beam damage, external disturbance, or lens aberrations.

The incoming probe current and the outgoing electron flow are almost equal during the observation of conductive samples. On the contrary when observing non-conductive samples, the total number of electrons emitted from the specimen and flowing through it is not equal to the number of the incident ones; in fact the negative charge of the primary electron beam charges the sample surface, changing its potential and causing image disturbance or abnormal contrast. Avoiding the charge-up phenomenon is possible by reducing the accelerating voltage and the beam current, coating the sample with a conducting layer in order to drain the charges to ground, neutralizing the induced charges by the ionization of a gas present in the specimen chamber (low vacuum or ESEM mode), or filtering out the portion of the affected signal from artifacts by changing the bias or the collecting geometry of the detector.

When a charged sample is imaged at an accelerating voltage lower than the surface potential, the Coulombian repulsion between the irradiating electrons and the sample surface sends the incoming electrons back in the SEM chamber, along a variety of paths depending on the local details of the electromagnetic fields. Hence the charged particles cannot reach the sample surface that acts as a mirror and "reflects" electrons somewhere else in the specimen chamber. In this way the electrons do not penetrate the sample; instead they travel in any direction and hit anywhere inside the specimen chamber generating SE and BSE. The SE and sometimes also the BSE are then collected by the ETD and/or BSE detectors. This phenomenon that at the beginning was considered as a charging artifact and originally termed "anomalous contrast" is now known as electron mirror effect or pseudo-mirror effect and gives rise to a mirrorless reflection, i.e., a reflection in the absence of a material mirror. The same phenomenon is present also in ion columns under certain conditions and is known as ion mirror effect. The mirroring, usually described in terms of something very close to what happens to photons hitting an optical mirror, is a big source of ambiguity that sometimes gives images neatly defined but completely unrelated to the surface of interest, such as all the equipment facing the interior of the specimen chamber instead of an image of the sample surface of interest as expected (Fig. 2.3); other times the obtained images represent a mix of the images of both the specimen chamber and the sample; the balance of the mix is influenced by the properties of the sample, the electromagnetic properties of the beam, the experimental setup, and the sequence of the operations. The resulting mirror image is not stationary and varies in time since the sample and the environment electromagnetic properties mainly associated to the charge distribution in the sample and its surroundings are strictly interconnected, and, due to the reiterate interventions of the operator, the mirroring ability fades out.

Despite the mirror effect being studied since the 1970s, so far it is still not completely understood; surely the mirror effect arises numerous issues regarding the interpretation of the acquired images and the possible mechanisms that determine the electron paths. Much effort has been devoted to avoiding the appearance of this effect since it affects the image quality. On the other hand, the charged particle reflection (or to better say deflection or scattering, since electrons do not move along linear paths as in the classical optical reflection) can be a possible analytical tool for imaging or obtaining information about sample dielectric properties (Milani et al. 2010; Milani et al. 2009; Belhaj et al. 2000; Cazaux 1999; Luk'yanov et al. 1974; Clarke and Stuart 1970).

Residual gas molecules in the specimen chamber or molecules derived from the sample can adhere to its surface. When the electron beam hits the sample, the matter redeposited on its surface suppresses the discharge of SE reducing the brightness, and dark images are therefore produced. This contamination phenomenon is bypassed reducing the gas molecules in the specimen chamber, therefore increasing the vacuum level.

Polymeric materials and biological samples can be damaged by particle beams causing both thermal and chemical modifications on the sample. The temperature increment depends on numerous factors such as accelerating voltage and intensity of the beam, observation area and time, specific heat, and heat conductivity of the

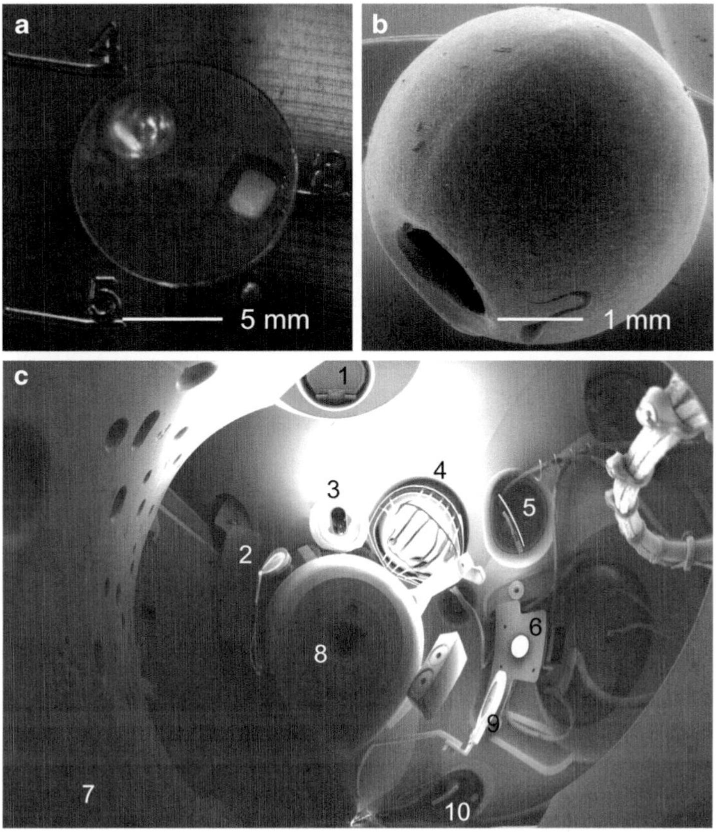

Fig. 2.3 Venus and mirror... effects: reminiscence of art in electron microscopy. The Venus effect is a common phenomenon in which "the mirror itself is used (deliberately or not) to lead us down the wrong path." The mirror effect is a type of Venus effect where the mirror is a self-built one so that the reflection of a subject that is not the target of the observation by the observer appears. The observer accepts this reflection as true, even though according to the laws of optics (also in the case of electron optics and electron microscopy), the reflection presented is totally unreliable and false (Bertamini et al. 2003). It can be said that electron mirror images act as mirrors in tales; they tell the truth but of places and times far away; in the same way mirror images give some reliable information, simply not the information one was looking for. The figure presents the results of a mirror imaging of a non-conductive plastic sphere. (**a**) Photograph of the sphere (diameter 4 mm) located on the top left of the SEM sample holder, acquired with the SEM internal navigation camera. (**b**) SEM image of the sphere obtained in low vacuum to prevent charging effects and the consequent anomalous contrast. (**c**) Mirror image of the sphere showing the internal part of the SEM chamber instead of the sample as expected. In (**c**) the mirror image is generated by the sphere that is epoxy glued onto an Aluminum stub to prevent the grounding of the sample; the stub is then loaded on the sample holder and the specimen chamber is evacuated to the high vacuum regime. The scanning of the electron beam (30 keV, ca. 10 nA) on the sphere for about 3 mins on an area of about $2 \times 1.5\ \text{mm}^2$ builds a negative electric field on top of the sample. The primary electrons generated by the imaging operations (2 keV, 130 pA) are bounced back in the SEM chamber and are then collected by the ETD. The generation of the mirror image appears to be a rather stable process even after imaging the sample many times for more than half an hour. The image is of good quality and provides the identification of several devices and details present in the chamber; anyway the image even with its metadata is not able to provide any spatial information of the observed objects. 1. EBSD Electron Backscatter Diffraction. 2. LFD Large Field Detector (SE for low vacuum). 3. EDS detector. 4. ETD. 5. STEM (retractable) detector in parking position. 6. CCD camera. 7. Stage table. 8. Final lens pole piece. 9. Heating stage shield. 10. BSE (retractable) detector in parking position

sample. To avoid beam damage, it is advisable to reduce the irradiating current, lower the accelerating voltage, and coat and cool the sample. Ion beams, due to the bigger mass of the constituent particles with respect to electrons, provoke major physical damages on the sample, taking away atoms and clusters, effectively modifying the shape of the specimen.

Other image abnormalities can be caused by inadequate preparation of the sample, improper insertion of the sample onto the specimen holder, vibrations, stray magnetic fields, low irradiation current, inadequate working distance and beam alignment, objective aperture contamination, microscope parameters not properly set, or lens aberrations. Typical problems regarding lenses are represented by spherical and chromatic aberrations, diffraction, and astigmatism. Spherical aberration is a distortion caused by different convergent positions of electrons passing near the lens center and electrons passing far from it. It can be reduced using a small aperture and a shorter focal distance. Chromatic aberration is caused by a variation of the wavelengths of electrons incident on the lens. This aberration can be reduced decreasing the energy spread of the electron beam. Diffraction can be reduced increasing the aperture diameter, but not enlarging it too much in order to not increase spherical and chromatic aberrations. Astigmatism occurs when the shape of the cross section of the electron beam (spot) due to asymmetrical magnetic fields, charge-up phenomenon, or contamination of the sample is not circular but elliptical, and, as a consequence, stretching and blurring effects are visible on the acquired images (Reimer 1997; Dykstra 1993).

Sample Preparation

Samples observed via electron microscopes are usually prepared following conventional methods. TEM and SEM investigations need different sample preparation protocols that comprise several steps and more than 24 hours are usually required to complete the procedure. Traditional preparation methods can be changed according to the sample under investigation (powders or fine particles, polymers and inorganic matter, crystals, metals, semiconductors, biological specimens), to the particular structure of interest (in biological samples: membranes, nuclei, vesicles, etc.), and to the type of electron microscope used; for some purposes chemical manipulation of the samples can be substituted by simple air drying as in the case of specific samples (bacterial and plant cells) observed via SEM or FIB/SEM microscopes in high vacuum. When asking people to think about what happens to a non-prepared cell put in the vacuum, it is easy to collect answers predicting either the explosion or the collapse of the cell; in both cases relevant damages and catastrophic alterations in the cellular morphology are expected. It is extremely common that people hardly believe what actually happens: cellular membranes and inner structures allow the cells to maintain their overall integrity notwithstanding the vacuum. This is an important issue for microscopists because it is possible to avoid preparation artifacts and to speed up the times that come between the collection of the sample and the actual observation, thus preserving the sample in its natural conditions, sometimes compromised as in forensic investigations, from xenochemical contamination.

Selecting the exact preparation method of a sample is a key step, useful to better interpret the acquired electron images. In the following part of this brief introduction to sample preparation are presented the standard procedures adopted for the observation of biological specimens via TEM and SEM.

For the observation via TEM, samples need to be ultrathin ($\leq$ 70 nm) and electron dense and must resist high vacuum and the electron bombardment. The conventional preparation of biological samples starts with a primary fixation (pre-fixation) with aldehydes. The fixation is important to stop the cellular processes and preserve the specimen in a state as close as possible to its natural one; hence a good fixative needs to quickly permeate the sample without creating fixation artifacts and must be irreversible. Glutaraldehyde thanks to its ability to fix proteins is the most common fixative used, usually at 2.5%. After the primary fixation the sample is washed with buffer and then fixed a second time (post-fixation) with Osmium tetroxide, a heavy metal able to fix lipids. Since Osmium tetroxide is electron dense, it is used not only as a fixative but also as a stain; in fact it improves the preservation of cell membranes and increases the contrast. After post-fixation the sample is washed once more with buffer and then undergoes a further fixation with tannic acid or uranyl acetate (conductive staining). This is an optional step important to avoid the sample charging. In order to replace the water contained in the sample preserving its morphological properties, the specimen is dehydrated with a gradual series of ethanol solutions. The alcohol is then replaced by epoxy or acrylic resin, through resin infiltration. Since biological specimens are soft, they need to be embedded in resin before cutting; the resulting specimen block is then trimmed in a pyramid-like shape and cut with a microtome equipped with a glass knife in semithin sections (1 μm). Semithin sectioning is for optical microscopy and to approach the point of interest (POI) in the pyramid-like shape. Once the POI is reached, the resin block is transferred to an ultramicrotome, sliced in thin sections (50–100 nm), and placed on a TEM grid. Sections are then stained with uranyl acetate or lead citrate and eventually can be observed in a TEM. Ultramicrotome cutting is not the only method of preparation of ultrathin sections, they can be cut using the ion beam of a FIB. In this case the cut is advantageously site specific with the possibility of choosing cutting planes not necessarily orthogonal to the sample' surface; the only disadvantage is represented by the contamination by Gallium ions implanted by the ion primary beam.

In order to avoid the appearance of artifacts deriving from the preparation techniques, several attempts to find new methods to preserve the sample in a condition close to its natural state have been made. One of the procedures is the freezing technique: there are several types of freezing methods, for instance the Tokuyasu one, excellent to preserve antigens, hence mostly used in immunology: the freeze cutting method which allows to directly observe fresh tissue; the freeze replica method used to investigate the cellular structures inside the cell membrane; and the cryo-embedding method in which a sample is rapidly frozen on a grid, embedded in ice, and directly observed.

For the observation in a SEM, dimensional restrictions are less compelling than in a TEM; samples usually need to be conductive and stable in the vacuum.

Specimens observed in SEMs can be prepared according to the conventional methods or not prepared at all. The standard preparation protocol of biological samples for the observation in SEMs starts with the same steps as for the TEM preparation method: primary fixation, washing with buffer, secondary fixation, washing, conductive staining, and dehydration with alcohol. At this point the alcohol must be removed from the sample; a transition fluid is required to not introduce surface tension or drying artifacts: hexamethyldisilazane (HMDS) and liquid CO_2 (critical point drying technique) are the most used. After the drying step the sample can be placed on a specimen stub and fixed with double-sided conductive tape or with conductive graphite paint. Conductive tape or paste is used not only to fix the sample to the specimen stub but especially to ground the sample, preventing its charge-up. Usually the sample undergoes a metal (Gold, Gold-Palladium, Platinum-Palladium, Platinum) or Carbon coating in order to make the sample surface conductive, to increase the production of SE, and to prevent damages to the specimen. The coating thickness is usually standard, but it can vary according to the magnification used for the observation, a thinner coating being better for high magnifications, a thicker one being better for low magnifications (Glauert and Lewis 2014; Ayache et al. 2010; Dykstra 1993).

Biomedical Applications of Electron Microscopy

In this section different electron microscopy techniques applied to different biological samples are presented to give a feeling of electron microscopy potential and limits.

An electron microscopy analysis was conducted on renal calculi collected from patients suffering from nephrolithiasis, a pathology caused by the presence of kidney stones and characterized by local inflammation, pain, and functional alterations of the organ. Microorganisms participating in the formation of renal concretions, such as *Escherichia coli*, remain in renal calculi for a long time, being at the base of septic complications. *E. coli* is a Gram-negative bacterium able to produce a biofilm, cause infectious processes, and provoke commissure formations which fix stones to kidneys. Electron microscopy analyses of concretions from patients suffering from nephrolithiasis, who underwent surgery or lithotripsy, showed the presence of collagen fibers on the stones (Fig. 2.4) suggesting the requirement of a proper surgical methodology for calculi removal, in order to prevent the infection from spreading and to avoid a relapse of nephrolithiasis (Didenko et al. 2014).

A FIB/SEM application relevant for environmental studies was conducted on a terrestrial isopod crustacean, *Porcellio scaber*. This organism is a detritophage, which accumulates metals (even those toxic for human beings); thus it is helpful to monitor the effects of human settlements on the terrain. Figure 2.5 shows SEM images of hepatopancreatic (digestive gland) S-cells of *P. scaber*, prepared according to standard SEM preparations. At different magnifications it is possible to observe the structure and components of the crustacean hepatopancreas, a hollow tube formed by S- and B-cells. S-cells store metals, in particular Copper and Zinc, in granules of homogenous material; B-cells store more loosely bound material in

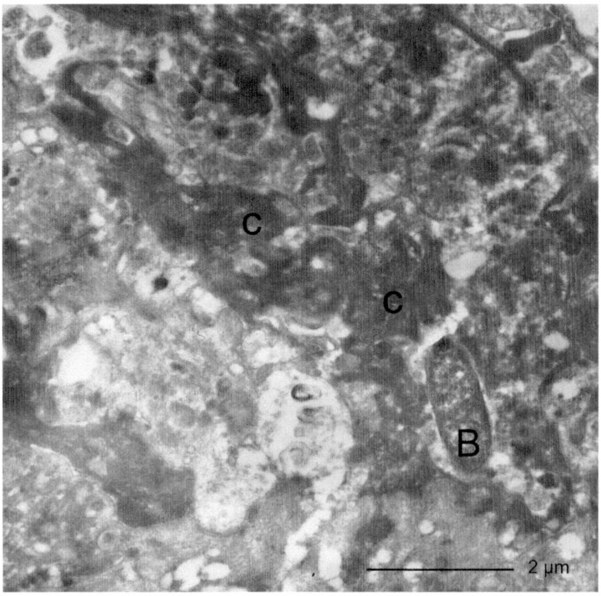

Fig. 2.4 TEM image of the surface of a renal calculus. The nephrolith is covered by a biofilm composed of Gram-positive and Gram-negative bacteria (B) localized between collagen fibers (C) and cell debris

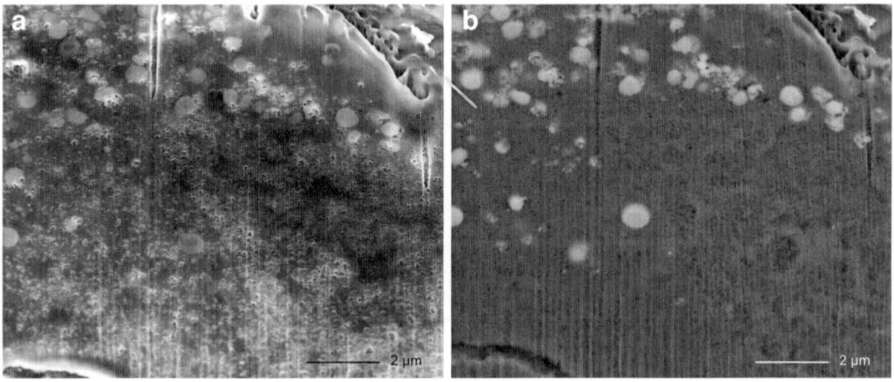

Fig. 2.5 SEM images of a region of a hepatopancreatic S-cell of *P. scaber* obtained after FIB ion milling with a FIB/SEM microscope. (**a**) SE image shows two portions of the borders of the cell. It is possible to evaluate the final quality of the surface after ion milling and polishing operations. The ion beam going from up to bottom creates a curtain effect that alters the ideal flatness of the surface. (**b**) BSE image; this technique highlights the homogenous bright areas, indicating the presence of aggregates of elements with higher atomic numbers (for instance metals)

which Iron and Calcium (fundamental for the exoskeleton of the organism) are present. Images were acquired, after ion milling operations, with different techniques: Fig. 2.5a is a SE image which is useful to characterize the morphology of the sample; Fig. 2.5b is a BSE image able to enhance the physicochemical differences in the area of interest. Due to the depth from which the signal is produced, the BSE image has a lower resolution than the SE one (Drobne et al. 2005).

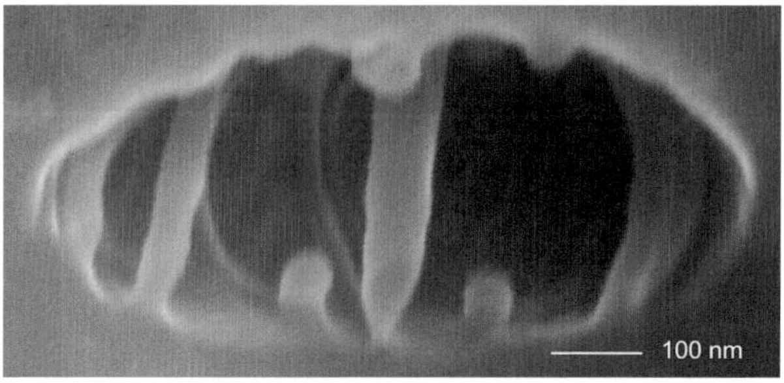

Fig. 2.6 SEM image of a *Saccharomyces cerevisiae*'s mitochondrion acquired with a FIB/SEM microscope after ion milling and polishing operations. In the image the inner and outer mitochondrion's membranes are clearly visible together with *cristae*. (Image by C. Savoia and M. Milani)

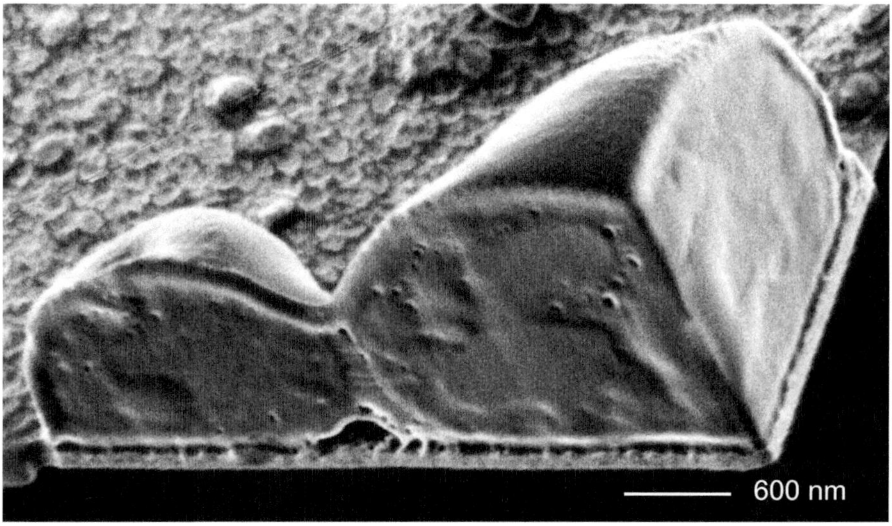

Fig. 2.7 Ion image (at 50 kV) of a section of an unprepared (air dried) *S. cerevisiae* cell with its budding acquired in a FIB after milling and polishing operations. The cell was cut along two orthogonal axes and an unexpected sharpness of the resulting corner that is comparable with the sharpness of the cut of the silicon substrate is observable. The external membranes together with the internal organelles and the budding junction are visible. (Image by H. Lezec and M. Milani)

Figures 2.6 and 2.7 show electron and ion microscopy images of unprepared (air dried) samples of *Saccharomyces cerevisiae*. Yeast cells maintain their structural integrity even in high vacuum; thus it is possible to perform operation of manipulation and imaging both with the ion and the electron beams. The ion milling has been used to expose sections of high quality at selected points without provoking structural collapse of the biological object under investigation; successively electron and ion images of the obtained sections have been acquired. The *in situ* manipulation and imaging of cells not fixed according to conventional methods proved the possibility to investigate morphological and ultrastructural details of unprepared samples (Milani et al. 2004).

References

Artsimovitch L, Loukianov S. Mouvement des particules chargées dans des champs électriques et magnétiques, MIR, 1975.

Ayache J, Beaunier L, Boumendil J, Ehret G, Laub D, editors. Sample preparation handbook for transmission electron microscopy. New York: Springer; 2010. https://doi.org/10.1007/978-0-387-98182-6.

Baird D, Shew A. Probing the history of scanning tunneling microscopy. In: Baird D, Nordmann A, Schummer J, editors. Discovering the nanoscale. Amsterdam: IOP Press; 2004.

Bassett GA, Menter JW, Pashley DW. Moiré patterns on electron micrographs, and their application to the study of dislocations in metals. Proc R Soc London A. 1958;246(1246):345–68.

Belhaj M, Jbara O, Odof S, Msellak K, Rau EI, Andrianov MV. An anomalous contrast in scanning electron microscopy of insulators: the pseudo-mirror effect. Scanning. 2000;22:352–6. https://doi.org/10.1002/sca.4950220603.

Bertamini M, Latto R, Spooner A. The Venus effect: people's understanding of mirror reflections in paintings. Perception. 2003;32(5):593–9. https://doi.org/10.1068/p3418.

Bogner A, Jouneau PH, Thollet G, Basset D, Gauthier C. A history of scanning electron microscopy developments: towards "wet-STEM" imaging. Micron. 2007;38(4):390–401. https://doi.org/10.1016/j.micron.2006.06.008.

Candia Carnevali D, Milani M. Electron and ion microscopy and micromanipulation: common principles and advanced methods in applied sciences. Proceedings of Summer School 2008 & 2009, Esculapio, Bologna, Italy, 2010.

Cazaux J. Some considerations on the secondary electron emission, δ, from e− irradiated insulators. J Appl Phys. 1999;85:1137–47. https://doi.org/10.1063/1.369239.

Chadwick J. The existence of a neutron. Proc R Soc London A. 1932;136:692–708. https://doi.org/10.1098/rspa.1932.0112.

Clarke DR, Stuart PR. An anomalous contrast effect in the scanning electron microscope. J Phys E. 1970;3:705.

De Broglie L. Recherches sur la théorie des Quanta (Doctoral dissertation, Migration-université en cours d'affectation), 1924.

Didenko LV, Tolordava ER, Perpanova TS, Shevlyagina NV, Borovaya TG, Romanova YM, Cazzaniga M, Curia R, Milani M, Savoia C, Tatti F. Electron microscopy investigation of urine stones suggests how to prevent post-operation septic complications in nephrolithiasis. J Appl Med Sci. 2014;3(4):19–34.

Drobne D, Milani M, Zrimec A, Zrimec MB, Tatti F, Draslar K. Focused ion beam/scanning electron microscopy studies of *Porcellio scaber* (Isopoda, Crustacea) digestive gland epithelium cells. Scanning. 2005;27(1):30–4. https://doi.org/10.1002/sca.4950270106.

Dykstra MJ. A manual of applied techniques for biological electron microscopy. New York: Springer; 1993.

Egerton RF. Physical principles of electron microscopy. Cham: Springer; 2016. https://doi.org/10.1007/978-3-319-39877-8.

Feynman RP. There's plenty of room at the bottom. Eng Sci. 1960;23:22–36.

Giannuzzi LA, editor. Introduction to focused ion beams: instrumentation, theory, techniques and practice. New York: Springer Science & Business Media; 2004.

Glauert AM, Lewis PR. Biological specimen preparation for transmission electron microscopy. Princeton: Princeton University Press; 2014. https://doi.org/10.1515/9781400865024.

Hawkes P. Recent advances in electron optics and electron microscopy. Ann Fonda Louis de Broglie. 2004;29(1):837–55.

Karlsson Rosenthal C. The beginning. Nat Cell Biol. 2009;11:S6. https://doi.org/10.1038/ncb1938.

Krumeich F. Properties of electrons, their interactions with matter and applications in electron microscopy, 2011. https://www.microscopy.ethz.ch/downloads/Interactions.pdf

Levey S. Leibniz on mathematics and the actually infinite division of matter. Philos Rev. 1998;107(1):49–96. https://doi.org/10.2307/2998315.

Liu J. Advances and applications of atomic-resolution scanning transmission electron microscopy. Microsc Microanal. 2021:1–53. https://doi.org/10.1017/S1431927621012125.

Luk'yanov AE, Spivak GV, Gvozdover RS. Mirror electron microscopy. Sov Phys Usp. 1974;16(4):529–52.

Mansfield E, Kaiser DL, Fujita D, Van de Voorde M. Metrology and standardization for nanotechnology: protocols and industrial innovations. Weinheim: Wiley; 2017.

Meinel C. Early seventeenth-century atomism: theory, epistemology, and the insufficiency of experiment. ISIS. 1988;79:68–103. https://doi.org/10.1086/354634.

Milani M, Abdul-Wahab HN, Abbood TH, Savoia C, Tatti F. "Rear window": looking at charged particles hitting a charged target in a FIB/SEM. In: Méndez-Vilas A, Díaz J, editors. Microscopy: science, technology, applications and education. Madrid: Formatex; 2010.

Milani M, Ballerini M, Batani D, Squadrini F, Cotelli F, Lora Lamia Donin C, Poletti G, Pozzi A, Eidmann K, Stead A, Lucchini G. High resolution microscopy techniques for the analysis of biological samples: a comparison. Eur Phys J Appl Phys. 2004;26(2):123–31. https://doi.org/10.1051/epjap:2004029.

Milani M, Drobne D, Drobne S, Tatti F. An atlas of FIB/SEM applications in soft materials and life sciences. Aracne, 2006.

Milani M, Savoia C, Bigoni D. Electron mirroring: control of electron transport and understanding of physical processes from SEM images. In: Proceedings of ITP2009. Interdisciplinary transport phenomena VI: fluid, thermal, biological, materials and space sciences, 2009.

O'Keefe MA. Seeing atoms with aberration-corrected sub-Ångström electron microscopy. Ultramicroscopy. 2008;108(3):196–209. https://doi.org/10.1016/j.ultramic.2007.07.009.

Orloff J, Swanson L, Utlaut M. High resolution focused ion beams: FIB and its applications. In: The physics of liquid metal ion sources and ion optics and their application to focused ion beam technology. Boston: Springer; 2003. https://doi.org/10.1007/978-1-4615-0765-9.

Reimer L. Transmission electron microscopy – physics of image formation and microanalysis. Berlin: Springer-Verlag; 1997. https://doi.org/10.1007/978-3-662-14824-2.

Rohde M. Microscopy. Methods Microbiol. 2011;38:61–100. https://doi.org/10.1016/B978-0-12-387730-7.00004-8.

Ruska E. The development of the electron microscope and of electron microscopy. Biosci Rep. 1987;7(8):607–29. https://doi.org/10.1007/BF01127674.

Rutherford E. Bakerian lecture: nuclear constitution of atoms. Proc R Soc London A. 1920;97:374–400. https://doi.org/10.1098/rspa.1920.0040.

Stokes DJ. Principles and practice of variable pressure/environmental scanning electron microscopy (VP-ESEM). Chichester: Wiley; 2008. https://doi.org/10.1002/9780470758731.

Thomson G, Reid A. Diffraction of cathode rays by a thin film. Nature. 1927;119:890. https://doi.org/10.1038/119890a0.

Thomson JJ. Cathode rays. Lond Edinb. 1897;44(269):293–316. https://doi.org/10.1080/14786449708621070.

Ul-Hamid A. A beginners' guide to scanning electron microscopy. Cham: Springer Nature; 2018. https://doi.org/10.1007/978-3-319-98482-7.

Volkert CA, Minor AM. Focused ion beam microscopy and micromachining. MRS Bull. 2007;32:389–99.

Wells OC, Gordon MS, Gignac LM. Past, present, and future of backscatter electron (BSE) imaging. In: Scanning microscopies 2012: advanced microscopy technologies for defense, homeland security, forensic, life, environmental, and industrial sciences, 2012. p. 8378. https://doi.org/10.1117/12.920001.

Williams DB, Carter CB. Transmission electron microscope. A textbook for materials science. Boston: Springer; 1996. https://doi.org/10.1007/978-1-4757-2519-3.

Nanoworld

The ability to reduce everything to simple fundamental laws does not imply the ability to start from those laws and reconstruct the universe.
Philip W. Anderson

History

The term "nanotechnology" was coined by Norio Taniguchi in 1974, but it was R. P. Feynman in December 1959 that actually founded the concept of nanotechnology. In his speech "There's Plenty of Room at the Bottom," Feynman talked about the possibility of manipulating and synthesizing matter at an atomic scale, handling and dismantling nanoparticles, working down from the macro- to the micro-scopic level, and arranging atoms the way one wanted, "the very atoms, all the way down!" (Feynman 1960). It was 1986 when the term nanotechnology was used again, this time by K. Eric Drexler in *Engines of Creation: The Coming Era of Nanotechnology*, who proposed the idea of using machines that working at the molecular scale could structure matter from the bottom up (Hulla et al. 2015).

Is "the lawn" what we see or do we see one grass plus one grass plus one grass…?
Italo Calvino

The task of nanosciences and nanotechnologies is the manipulation of the elementary constituents of matter, atom by atom, from functionalized to specific applications, sometimes overlapping macromolecular chemistry. The bottom-up and the top-down approaches are two different ways of building nanostructures: the bottom-up method exploits the atomic and molecular forces, i.e., electrodynamics at the fundamental (quantum) level leading to aggregation and separation of atoms in order to organize and assemble molecules into nanomaterials, whereas with the top-down technique nanostructures are carved from bulk materials into the desired shapes. During the manufacturing of nanostructures the top-down approach is preferred rather than the

M. Milani et al., *Bacterial Degradation of Organic and Inorganic Materials*,
https://doi.org/10.1007/978-3-031-26949-3_3

bottom-up, but often the two processes are combined. An example is provided by photonics with top-down optical antennas fabricated by FIB milling, and bottom-up optical antennas consisting of Gold nanoparticles supported by a dielectric tip (Manoccio et al. 2021; Joachim 2020; Novotny and Van Hulst 2011; Magni et al. 2006).

Between 1981 and 1991, with the STM and the AFM, what people thought of as a visionary plot proved to be a feasible project; the developments in the fields of transmission and scanning electron microscopy and the consequent expansion of the range of action of collimated electron and ion beams to more heterogeneous samples led nanotechnology to a phase of rapid growth. During these years, the ability of imaging surfaces at atomic scale propelled a great development even in interface and colloid science that, although already existing, had never been associated to nanotechnologies until then. That was the time of the discovery of fullerenes, of multi- and single-walled Carbon nanotubes, and of the comprehension of the new properties of these nanomaterials (Lin et al. 2007).

Since then nanotechnologies became part of a lot of areas, with an increasing role of nanoparticles and nanodevices in physical, chemical, biological, biomedical, and pharmaceutical sectors, as well as in electronics and energy harvesting applications, and in the ecological, commercial, and industrial field (Khan et al. 2019; Jeevanandam et al. 2018). This advancement on nanoscale technologies though did not involve any atomic matter handling because the process would have been too slow, faulty, and not well controlled, but merely the addition of nanosized particles to bulk structures (Bittner et al. 2013).

> *The direct observation of resolved single atoms, atom-by-atom cluster formation and the crystallization process for nanoscopic crystals is still an experimental frontier.*
> *Nicolas P.E. Barry*

Nowadays it is nearly impossible not to come into contact with a nanomaterial or with something whose production did not involve at some stage the use of nanomaterials and related technologies. If for some items, such as cosmetics, pharmaceuticals, electronics, cleaning products, pesticides, or fuel additives, it is simple or just more common to associate the presence of nanosized materials, in others, for instance toys, food packaging, car accessories, and sporting goods, the presence of some sort of nanocomponent is not so evident.

Properties of Nanoparticles

Nanoparticles are among the most studied nanomaterials of the last several decades, but despite this a definition internationally shared of the term "nanoparticle" does not exist. According to the International Organization for Standardization (ISO) a nanoparticle is a discrete nano-object where all three dimensions are less than 100 nm. The US Food and Drug Administration and the Commission of the European Union define nanoparticles as materials that have at least one dimension in the 1×10^{-9} and 1×10^{-7} m range (Jeevanandam et al. 2018). The absence of a shared definition has been criticized by scientists due the consequent impossibility to have a proper regulatory action; curiously also the hypothetical presence of a shared definition has not found favor with other groups of scholars who claimed that a single definition while serving a certain purpose well, could fail to capture other important aspects, especially when trying to assess the risk of a nanomaterial. In this case as well a proper regulatory action would be prevented, thus representing an obstacle to the scientific and technological advancement (Wigger et al. 2018; Boholm and Arvidsson 2016).

Nanoparticles are classified according to several criteria: material-based categories (Carbon-, inorganic-, organic-, or composite-based), morphology (spherical, rod-like, irregular, crystal-like), origin (natural or synthetic), and source (incidental, i.e., a byproduct of industrial processes or of natural events such as forest fires, dust storms, or volcanic eruptions; engineered, i.e., especially manufactured to have peculiar features; natural, i.e., found in plants, animals, and human organisms) (Jeevanandam et al. 2018). Due to their size also nanobacteria and viruses can be considered as natural nanomaterials and used for nanotechnology applications, actually abstracting them from their main infective role. Having precise nanoscale structure and dimensions, viruses are seen as well-ordered materials; viral capsids have internal cavities accessible to small molecules, thus providing an opportunity for the assembly and packaging of cargoes; moreover viruses are biologically and chemically addressable and their properties can be influenced, manipulating the viral genome (Ryu 2017; Bittner et al. 2013).

The wide use of nanomaterials in a lot of applications is due to the peculiar properties of nanoparticles such as large surface/volume ratio, size-dependent optical properties, chemical reactivity, magnetic properties, thermal conductivity, and several mechanic parameters as elastic modulus, hardness, stress, strain, adhesion, and friction. Nanoparticles show unique characteristics even compared to the same material at a bigger scale: for a bulk material the number of atoms laid out at the surface is not relevant in comparison to the total quantity of atoms of the whole material; on the contrary at the nanoscale the surface/volume ratio changes and the number of atoms arranged at the surface of the nanoparticle becomes very significant. The consequent augmented reactivity, enhanced mobility, and ability to penetrate barriers make nanoparticles more suitable for nanofabrication and nanomanufacturing processes (Khan et al. 2019). Depending on their size, surface/volume ratio, structural and optical characterization nanoparticles can be dispersed in aerosols and suspended in liquids or emulsions.

The particles, at the nanoscale, interact with each other through van der Waals forces, polar, electrostatic, and covalent interactions (different manifestations at the quantum level of the fundamental electromagnetic interaction between atoms and molecules). The presence of chemical structures (even water molecules) in the medium in which nanoparticles are dispersed can affect their surface charge, altering it and inducing modifications in the nanoparticles' behavior, for instance enhancing or hindering their aggregation or coagulation. The details of nanoparticle-nanoparticle interactions driven by attractive or repulsive forces are fundamental for the understanding and controlling of the mechanisms occurring at the nanoscale.

So far it has been demonstrated that the cellular uptake of nanoparticles depends on the surface charge and is more influenced by the nanoparticle aspect ratio, i.e. the proportion between width and height, than by their size and shape. Aggregation of nanoparticles is another important issue in the nanoparticle-cell uptake. Usually, studies take into account only the cellular uptake of single nanoparticles, but in special formulations nanoparticles are in the form of aggregates (since the very beginning or as a consequence of the aggregation forced by the dispersion medium) and their interactions with cells and subsequent uptake need to be clarified (Adjei et al. 2014).

Though under investigation for years nanoparticles' properties still need to be studied, especially in fields such as foods, nanomedicine, and life and environmental sciences, where nanomaterials are extensively used with some awareness but without the full control of the interactions that nanoparticles could have with biological components and the environment. Through experimental studies nanotoxicology, the branch of toxicology related to nano-objects, has the task to assess which adverse biological responses, from health issues to toxicity, nanoparticles could provoke (Fadeel 2019; Angioletti-Uberti 2017; Forest et al. 2015).

Nanoparticles and the Quantum Realm

Characteristics shown by nanomaterials and the differences of mechanical, thermal, and catalytic properties compared to those of bulk materials require a quantum mechanical approach: quantum dynamics and effects are to be considered whenever the nanoscale is reached.

> *Atoms on a small scale behave like nothing on a large scale,*
> *for they satisfy the laws of quantum mechanics.*
> *Richard P. Feynman*

The first issue to be questioned is whether a nano-object has to be seen as the result of the division of a macroscopic one or from the addition of several objects at atomic scale. During the buildup of nanostructures, when passing from single atoms to sets of organized ones (from single atoms to nanoparticles or crystals), novel physical or chemical properties, drastically different from those of the corresponding bulk material, appear. The bottom-up approach, where the size pushes and pulls the new object in and out the classical and quantum domain, highlights the limit of quantum mechanical descriptions and the necessity to introduce a new discipline of statistical nature, like thermodynamics, able to handle the dynamics of collective systems made of many identical or nearly identical particles (Nicolis and Prigogine 1977). The behavior of aggregates of elementary particles cannot be understood, merely extrapolating the properties of small groups of particles, because at each level of complexity new collective properties appear. Synergetics, an interdisciplinary science that analyzes and compares the dynamics of order and collective phenomena in systems made of many identical interacting bodies, has the task of finding the underlying common mechanisms governing these cooperative effects

(Haken 1978). It is necessary therefore to define in a non-ambiguous way the concept of order. From this investigation emergent properties arise; they are higher-level properties which cannot be deduced from or explained by the properties of the lower level since complex systems have the ability to exhibit characteristics distant to those of their constituents. An example of emergent properties is represented by the liquid-solid transition of water at thermal equilibrium, with the multifaceted resulting structures; an illustration is represented by snowflakes, where low amounts (hundreds) of water molecules arrange in a multiplicity of spatially symmetrical forms (crystals) with fractal patterns of high artistic content (Anderson and Stein 1987; Anderson 1972; Hooke 1665).

The emergence of properties in physical phenomena, like rigidity in bulk matter or coherence in electromagnetic fields, supports the fact that quantity brings to quality differentiation in the broken symmetry or symmetry breakdown approach, in which small fluctuations strongly influence the system's dynamics, determining its transient or stationary state. The system, when subjected to transformations such as translation, rotation, or reflection, undergoes phase transitions, of the first or the second order, where states of reduced symmetry develop (Anderson 1972).

Rigidity looks like an intuitive simply measurable property (order parameter) emerging in a liquid-solid phase transition; rigidity is a collective name used in different phase transitions for several other properties, among which permanent magnetism and superconductivity, the latter being a phase of matter belonging to the class of Bose-Einstein condensates. This phase transition, characterized by an observable at macroscopic scale (order parameter), takes place when an ensemble of elementary constituents, as a consequence of their intrinsic or buildup quantum statistical (bosonic) properties, condensate in the lowest energy state (Anderson et al. 1995; Bose 1924; Einstein 1924). Anyway, rigidity appears as a property clearly understandable on the basis of daily experience, but it can have astonishingly debatable issues for instance when one moves from the properties of a diamond under mechanical stress to the properties of a biological sample that is milled by a FIB.

Philip W. Anderson claimed that a broken symmetry theory does not exist for systems far from the thermal equilibrium (dissipative structures), although he described several experiments that led to very similar types of symmetry breakdowns in the dissipative case. A dissipative structure appears in open, non-isolated systems which exchange energy with the environment. Unlike systems at equilibrium, a dissipative structure under suitable conditions can exchange energy in a highly asymmetric way. This asymmetry is not represented only by the direction of the energy exchange but also by its order: systems at equilibrium absorb disordered or thermal energy and reemit it in a disordered form (chaotic or black body radiation emission), whereas systems far from equilibrium, seemingly violating the second principle of thermodynamics, absorb disordered energy and reemit it "orderly" (coherent radiation). Anderson agreed to the existence of a similarity between systems at thermal equilibrium and systems far from the thermal equilibrium: the emergence of spatial variations from a homogeneous background (Anderson and

Stein 1987). There is a well-developed theory for the equilibrium case involving the order parameter concept, which leads to a strong correlation of the order parameter over macroscopic distances in the broken symmetry phase. Notwithstanding the assertions of Anderson, stable dissipative structures exist. The example of a dissipative structure *par excellence* is the LASER (light amplification by stimulated emission of radiation). Symmetry-breaking effects in lasers, systems well investigated from both the theoretical and experimental side, drive the stability of dissipative structures with the buildup of an order parameter (coherence length over macroscopic distances or coherence time over macroscopic time intervals) in the phase with reduced symmetry (coherent phase). Dissipativity is guaranteed by the fact that a laser is an open system where energy flows out towards the environment and is simultaneously acquired from the outside in the form of pumping energy or of feedback (Haken 1978).

The feedback can be represented as a further coupling between the laser and the environment, typically realized by one or more additional mirrors (the feedback source) which "reflect back" some of the emitted laser photons and define one or more external cavities. These photons have in turn a double activity: some of them simply enter the active cavity and fly across it, changing the total amount of photons inside it without affecting the active material in any way; another part interacts with the active elements in the cavity: carriers in semiconductor lasers, atoms and molecules in solid state, or gas lasers (Svelto 2010).

The role of feedback is intriguing since the total amount of the fed energy is commonly considered to be more important as the energy amount is increased, but the amount of the supplied energy is not the only factor able to transform the considered system in a highly nonlinear device where the effects are not driven simply from the amount of energy redirected back into the system. Actually small feedback fields produce the "observed" breakdown of the left-right symmetry in the laser (front-, rear-laser facets taking as reference semiconductor chip-on-carrier lasers). The reaction field introduces effects which span the whole volume of the system; even a relatively low amount of feedback from an external reflector under certain conditions creates instability driving the device into the so-called "coherence collapse" regime, characterized by tremendous noise. Finite volume effects set a threshold for the feedback field, above which symmetry breakdown appears (Brivio et al. 1993).

The same reasoning for the space domain can be equally applied to the time domain, where coherence is associated to time.

Systems far from the thermal equilibrium were systematically investigated by Grégoire Nicolis and Ilya R. Prigogine resulting in the "Brusselator," a theoretical model specially devoted to enzyme dynamics (Nicolis and Prigogine 1977). Substantially the Brusselator is equivalent to the laser model proposed by Hermann Haken and used to investigate "non-physical systems" such those found in biology, chemistry, psychology, neuroscience, and sociology (Haken 1978). Like Haken, several scientists chose to use the laser's emergent property of inanimate matter as a model for the nature and the origin of life, or for communication among individuals (Alodjants 2022).

The basic mechanism leading to ordered states is well described by the laser model and is due to nonlinear interactions between the microsystems (atoms) and the elementary components of the macrosystem. Overcoming the threshold value resulting from the balance between gain and losses, these interactions lead to long-range correlations among microsystems, to the instability of disordered states at equilibrium (i.e., without external energy pumping), and to the consequent appearance of ordered states (the emergence of correlations among the emitted photons). This is an operational definition of coherence, a powerful concept unfortunately often misused in many scientific and non-scientific fields.

The competition between gain and losses, which expresses the competition between two populations (similar to the Lotka-Volterra model), is the base of the so-called semiclassical approach and is an example of the passage from microphysics (photons, discrete variables) to macrophysics (electromagnetic field, continuous variables), also connected to phase and order-disorder transitions (Caldirola 1974).

The investigation of the "thermodynamic" aspects of the laser proves that this macroscopic system derives its characteristics from quantum physics. Therefore, in order to investigate the elements of the nanoworld (nanoparticles), it is necessary to mix quantum and classical physics, in the frame of synergetics.

The discovery of the microworld fundamental laws together with relativity represents the major progress of physics in the twentieth century. Despite its success, quantum mechanics is still misunderstood, especially in its philosophical implications, and often the source of endless disputes. These conceptual uncertainty and confusion in the scientists' minds are even more worrying for physics beginners and scholars investigating the microworld: "the students, emerging from their institutional courses of non-relativistic quantum mechanics, without exceptions showed how uncertain and uneasy their feeling was about a physical theory which is more than seventy years old, and permeates large sections of modern technology" (Preparata 2002). Generally the boundary between micro- and macro-phenomena is quite conditional and flexible. This passage is at stake since two different languages are commonly used to describe the micro- and the macro-world, with quantum physics acting as transient and intermediate language with the expectation of a future more homologous one. Moreover classical concepts are frequently found useful when considering microphenomena, whereas quantum mechanical ideas help in understanding the macro ones. The laser model and synergetics in general, as well as superconductivity and its associated phenomenology, are a bridge between both the micro- and macro-world and the quantum and classical language.

Hence in order to investigate the nanoworld, its elementary particles (nanoparticles), and their interactions with living matter, macroworld laws are not enough and they must be necessarily integrated with the often neglected microworld (quantum) ones, where nanoparticles are in general handled as ideal geometrical dots (without any reference to shape and size). In the case of normal fluids or solids the physical behavior of the macroscopic object can be described by the common classical mechanics language. But when dealing with systems that are a collection of a large (but not necessarily huge) number of atoms (elementary constituents) and with complex structures (for instance large polymers or macromolecules, nanoparticles), called intermediate systems, usual thermodynamic variables are no longer suitable to describe them and their behavior. As a consequence it becomes absolutely indispensable to describe "large" systems' properties in connection with their atomic, elementary structure and their being part of collective systems using quantum macro dynamics and synergetics.

The laser is a good model to discuss order-disorder transitions, the relations between order and disorder, and the very concept of order in a lot of non-physical systems since, although lasers range from quantum-dot to football-field size and utilize materials from free electrons to solids, the underlying operating scheme is always the same: energy sources convert input energy into light or, in general, electromagnetic waves. In the case of the laser (which for modeling necessity can be seen as a collection of atoms or electrical dipoles), the input or pump energy can assume different forms, with optical and electrical energy being the most common. In a conventional (incoherent) light source like a light bulb, a LED (light-emitting diode), or a star, each atom excited by an input pump energy emits a single photon at random times and directions. The radiation produced by all the atoms propagates through the whole space in all directions with the spread of wavelengths and obviously no interrelationships among individual photons (spontaneous emission). Albert Einstein predicted that excited atoms, when suitably interacting with photons, could release the stored energy converting it into light by a (resonant) process called stimulated emission. In this process the incoming photon, with definite properties related to the nature of atoms, emerges from the atom with a twin which has identical wavelength, direction, phase, and polarization; hence the stimulated emission is multiplicative, and the process can be iterated and leads to light "amplification." The emitted laser light can be detected when a positive balance between optical gain (depending on the number of excited atoms and on the energy pumped into the system) and optical losses (depending on the properties and shape of the active medium, i.e. the collection of atoms) is achieved (Webb 1972).

Classification and Description of Order and Disorder from Abstract Models

The discussion of ordered and disordered state and the transitions among them compels the investigator to become acquainted with the fact that in nature order and disorder are not the exact opposite of one another; instead they present a large spectrum of characteristics. What in common language is considered as a simple and unique feature, named order or disorder, reveals to be a nebulous concept with many facets often contradictory when discussed by different investigators; this can be seen for instance in some book titles as "*Obsessive-Compulsive Disorder and Uncertainty: Struggling with a Shadow of a Doubt*" or "*Harmonies of Disorder*" (Marcus and Tuber 2021; Montagnini 2017). The terms "order" and "disorder" are more than simple words; they represent a category described by an ensemble of roughly equivalent synonyms, although this relation is quite superficial and in a lot of cases misleading.

Science is characterized by a univocal correlation between the verbal description of a phenomenon or object and the presence of the phenomenon or object itself. Fundamental for the description/narration is the use of synonyms that both require and promote intrinsic, personal, subjective evaluations.

> *"Diversae quippe voces quae grammatico oculo perspectae synonymae sunt, eae nobis alia atque alia vel saltem aliter atque aliter significant."*
> *"For the different expressions which, when seen by the eye of the grammarian, are synonymous, theyor another, or at least in one way or another."*
> *Giordano Bruno*

Since they refer to the same meaning, synonyms are often used in an undifferentiated way, but people should choose which synonym to use depending on the shade of meaning they want to convey or the spatial and temporal perspective they intend to signify. In science the use of synonyms is often widely criticized since they could generate confusion and incertitude.

Associated to phase transitions, beyond the term order there is also the term crisis, obviously related to critical phenomena. Crisis can be defined in multiple ways, as the result of a trial or also as judgment (Merriam-Webster n.d.); both these meanings require the presence of alternatives and the existence of a question that of course implies an underlying doubt or uncertainty.

> *"Temptat enim dubiam mentem rationis egestas."*
> *"Our minds, disturbed by their inadequate knowledge of the truth, are uncertain."*
> *Titus Lucretius Carus*

An important example of terms often contextualized through controversial subjective descriptions is the pair order-disorder; these two words are strictly correlated in science, history, and literature as the Greek mythological pair cosmos-chaos.

Cosmos, meaning an ordered being, finds its roots in the Greek word *κόσμος* with which the ancient Greek referred to the universe and its ordered elegance. The same root is present in another common word: cosmetics. In this case the subjectivity of the beauty seen as elegant order is underlined. The same subjectivity is manifest in the graphs in which arrays of elementary objects, like the Greek capital letter Γ, are represented through different grades of order (Fig. 1). Order and elegance are a characteristic of any picture often depending on the magnification used to look at it.

Chaos is an old and evocative term derived from the Greek χάος that means chasm, emptiness, and vast void. In Greek mythology and philosophy chaos is always in contraposition with cosmos.

The revival of the term chaos is connected to the problem of weather forecasts and dates back to 1963 when Edward N. Lorenz stated the "chaos theory," the theory of systems that depending on small fluctuations of their initial state change their behavior in a fundamental way, leading to an exponentially growing uncertainty in the predictions of the future (Haken 1978; Lorenz 1963). However, the chaotic behavior, contrary to the common sense, does not always have catastrophic consequences; several systems maintain a predictable evolution even though slightly affected by a superimposed chaotic noise (Malvaldi and Leporini 2014; Ruelle 1994).

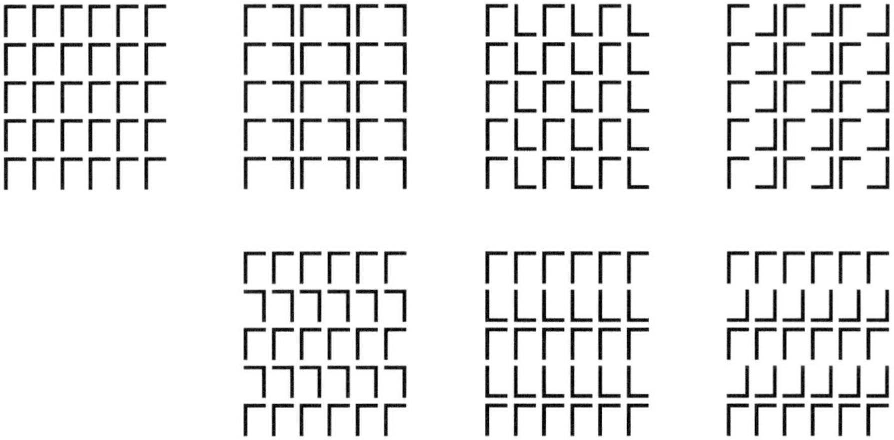

Fig. 1 Translation, mirror, and rotation symmetry can generate large objects from elementary units. This scheme is often present in artistic forms such as Alhambra mosaics (fourteenth century) and Maurits C. Escher's pictures. The elements of the figure graphically show how a set of elementary units can be ordered and consequently perceived in one of the three ways: individual (unitary element), group (subset), and global (set, ensemble) going from homogeneous distribution of elementary units to the perception of wave-like behaviors of the total surface

Usually it is assumed that order is preferable to disorder that, on the contrary, should be avoided; this bias is maximally evident in politics and for instance in psychology where mental disorders have a clear negative value.

The strict connection between order and disorder that highlights the subjectivity of the way in which such terms are used finds an historical representation in the sentence pronounced by Marc Caussidière, the prefect of police of Paris, during the French revolutionary events of 1848: "Il faut faire de l'ordre avec de désordre" that can be translated as "Disorder is at the base of order" or "You have to create disorder to obtain order" with some resemblance to Monte Carlo computational models (Grandjonc et al. 2009).

This sentence points out the importance of the environment in which the problem is contextualized; in particular the feature of irreversibility, typically associated to the concept of disorder, is directly related to the fact that the system under exam is either an open or a closed one (Heurtebise 2008).

Caussidière's political vision implies an open system and as a consequence the possibility of a driven organization of individual elements in contrast to a non-spontaneous or self-organized system. In other words, Caussidière's approach is a model of a thermodynamical phase transition driven by external forces whose effects depend on the flexibility and the mobility of individuals, the elementary units of the system influenced by the individual education. In physics features of rigidity, whose opposite is given by flexibility or unbreakability, are represented by the behavior of an ensemble of magnetic dipoles under the effect of a static external magnetic field where single units are identified by a three-dimensional vector; on the other hand an example of a set of single units is given by an ensemble of electrical dipoles under the effect of a static external electric field. In this case single units, made of tightly or loosely connected positive and negative charges, are subjected to orientation, deformation, and translation or to a mix of the previous effects. This kind of processes driven by external macroscopic fields, *de facto* coherent, is in competition with self-organizing processes induced by incoherent fields.

Another physical example, represented by the Cooper pair shows how a macroscopic field, without changing the individual electron behavior makes unusual properties on a different spatial scale emerge under suitable conditions. The dynamics of electrons is changed since they attract reciprocally to build a new elementary unit with completely different features leading to superconductivity (Cooper 1956). Cooper pairs can flow from one superconductor to another one (Josephson's tunneling in Josephson's junctions): a phase correlation (responsible for order) is obtained between the two superconductors that are separated by a layer originally disordered, and the long-range order is "transmitted" across the boundary/barrier (Josephson 1974; Josephson 1962). The whole system of the two superconductors, separated by a thin (1 nm) barrier, behaves as a single superconductor. Unlike ordinary superconductivity, this phenomenon is often called "weak superconductivity" because of the much lower value of the physical parameters involved, thus representing a clear example of different levels in the ordering process (Anderson 1964).

The complexity of the relation order-disorder is an important issue not only in the political or scientific fields but also in literature as shown, for example, in William Shakespeare's tragedies, where the order-disorder transition can be ruled by ambiguous and redundant language. A general pattern characterizes most of Shakespeare's works: in an ordered environment character's flaws are responsible for a disordered situation, the society suffers from this disorder, and in the end order is somehow reinstated. In Shakespeare's *Macbeth* the Witches, the "imperfect speakers," play with language, distorting it and communicating through inversions or unusual word order, thus generating ambiguity (El-Khazri 2015; Shakespeare 2004).

The terms order and disorder are multifaceted; they can be characterized in different ways, not always correlated with each other. This is well illustrated by the observations of Giacomo Leopardi and Milan Kundera regarding terms such as "infinite" and "nostalgia" and their translations into different languages (Kundera 2003; Leopardi 1917), at the basis of correlations and conflicts in the so-called two cultures (Snow 1959). Within the same language the role of synonyms is fundamental (for our purpose emphasis will be placed on the terms "picture" and "order"). In semantics, synonymy indicates the relationship between two lexemes with the same meaning, though admitting particular or stylistic characteristics and differentiated values (Fig. 2). The term synonym has itself some synonyms like analogous, equivalent, correspondent and affine, and the set of all these terms is often characterized by ambiguity, irrelevancy, or redundancy.

The considerations made so far, which touched different fields such as science, literature, politics, and art, aim to lead to a humanistic and scientific synergic approach to knowledge.

Fig. 2 One of the possible circular spatial dispositions of synonyms of terms frequently used in this manuscript; words normally used in fields distinct and distant from one another are placed side by side. The figure shows that the spatial distribution of the words, usually neglected or represented in a form of a rectangular table, is a good stimulus for imagination and helps to establish long-range connections among terms

What kind of relation does it exist between knowledge and reality? According to scientific realism, "the aim of science is to have true theories about the world" and science aims to an absolute and objective truth. Everyday knowledge as well as the scientific one, if it is possible to identify different areas in knowledge, arises from possessed cognitive faculties such as perceptual abilities, concept formation, deductive and non-deductive reasoning, hypothesis formation, and evaluation about objects and properties. All these faculties play a relevant role when doing science, but at the same time they are subjected to limitations being inevitably subjective qualities. This leads to the problem of the characterization of knowledge as absolute or objective, as well as the term truth, which shares lots of features with the word order. This is confirmed by the existence of "competing scientific theories that are independently confirmable but incompatible, and hence cannot be all true," as it happens for instance for quantum mechanics and relativity (Einstein's theory of gravitation) (Cellucci 2015).

"Non intelligimus, nisi phantasmata speculemur."
"We do not acquire experience, unless we explore "phantasmata"."
(imaginary representations, figures and similes).
Giordano Bruno

Reality both for science and life can be summarized by the sentence "people don't have a single integrated representation of complex issues… but rather rely on a patchwork of (sometimes disconnected or inconsistent) representations and can (without realizing it) dynamically shift between them when cued in context" (Thibodeau and Boroditsky 2011).

The different types of representation are correlated, as it happens for synonyms that are linked by a binary transitive property. After multiple readings of Fig. 1 it is evident that the observer establishes connections between order and disorder introducing extremely mobile and fluctuating boundaries that lead to the appearance of patterns and structures. In an analogous way the set of the terms illustrated in Fig. 2 can be organized according to different types of sorting as in Fig. 3, where the words

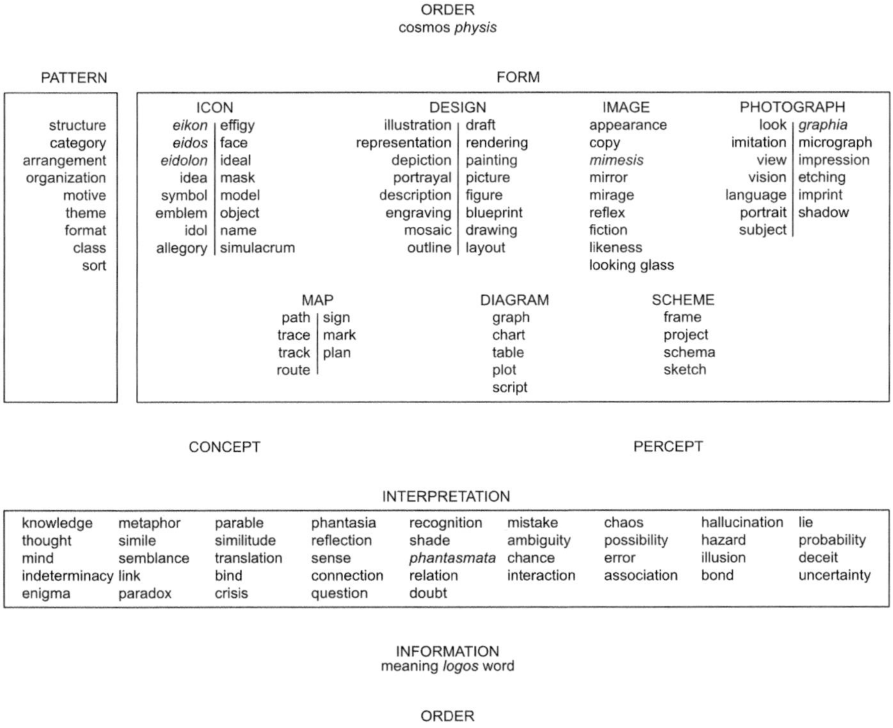

Fig. 3 The spatial disposition of synonyms in Fig. 2 can be represented as a tree diagram looping on itself (from order to order) in such a way that "roots" and "branches" can be the representation of two different worlds connected to each other by a multiplicity of paths. "Roots" and "branches" can also be a double representation of the same world (two-dimensional world if the existence of a third dimension is not admitted—or three-dimensional if the exploration of a third dimension is allowed)

were analyzed as a combination of several terms and only afterwards emerging arrangements were figured out.

In science and art, as well as in everyday life, disorder appears as a universal feature, inevitable and for some people even annoying and destabilizing. Order is brought or restored by *logos*, from Greek λόγος word, concept, idea, reason, and logic. Recalling the connection/boundary between order and disorder, it is important to emphasize that the *logos* is the element that helps to discern cosmos from chaos, exactly as the order parameter, associated to the slaving principle, differentiates two phases in a phase transition (Oakes 2014; Haken 1978).

The importance of *logos* and its necessary expression in the conflict between order and disorder is well represented in communication by the competition between signal (order) and noise (disorder), where the latter has a negative role in the generation and transmission of comprehensible information.

One of the most interesting aspects of the world is that
it can be considered to be made up of patterns.
A pattern is essentially an arrangement.
It is characterized by the order of the elements of which it is made
rather than by the intrinsic nature of these elements.
Norbert Wiener

References

Adjei IM, Sharma B, Labhasetwar V. Nanoparticles: cellular uptake and cytotoxicity. In: Capco D, Chen Y, editors. Nanomaterial. Advances in experimental medicine and biology. Dordrecht: Springer; 2014. p. 73–91. https://doi.org/10.1007/978-94-017-8739-0_5.

Alodjants AP, Bazhenov AY, Khrennikov AY, Bukhanovsky AV. Mean-field theory of social laser abstract. Sci Rep. 2022;12(1). https://doi.org/10.1038/s41598-022-12327-w

Anderson MH, Ensher JR, Matthews MR, Wieman CE, Cornell EA. Observation of Bose-Einstein condensation in a dilute atomic vapor. Science. 1995;269(5221):198–201. https://doi.org/10.1126/science.269.5221.198.

Anderson PW. Special effects in superconductivity. In: Caianiello ER, editor. Lectures on the many-body problem, vol. 2. Academic Press; 1964. p. 113.

Anderson PW. More is different. Broken symmetry and the nature of the hierarchical structure of science. Science. 1972;177(4047):393–6. https://doi.org/10.1126/science.177.4047.393.

Anderson PW, Stein DL. Broken symmetry, emergent properties, dissipative structures, life. In: Yates FE, Garfinkel A, Walter DO, Yates GB, editors. Self-Organizing Systems. Boston, MA: Springer; 1987. p. 445–57. https://doi.org/10.1007/978-1-4613-0883-6_24.

Angioletti-Uberti S. Theory, simulations and the design of functionalized nanoparticles for biomedical applications: a soft matter perspective. Computl Mater. 2017;3:48. https://doi.org/10.1038/s41524-017-0050-y.

Bittner AM, Alonso JM, Górzny MŁ, Wege C. Nanoscale science and technology with plant viruses and bacteriophages. In: Mateu M, editor. Structure and physics of viruses, Subcellular biochemistry, vol. 68. Dordrecht: Springer; 2013. https://doi.org/10.1007/978-94-007-6552-8_22.

Boholm M, Arvidsson R. A definition framework for the terms nanomaterial and nanoparticle. NanoEthics. 2016;10:25–40. https://doi.org/10.1007/s11569-015-0249-7.

Bose SN. Plancks Gesetz und Lichtquantenhypothese. Z Phys. 1924;26:178–81. https://doi.org/10.1007/BF01327326.

Brivio F, Grassia F, Graziano E, Mazzoleni S, Milani M, Vitiello G. Optical feedback induced mesoscopic effects in semiconductor lasers. Phys Lett A. 1993;175(1):5–10. https://doi.org/10.1016/0375-9601(93)90844-P.

Caldirola P. Dalla microfisica alla macrofisica. Edizioni scientifiche e tecniche Mondadori; 1974.

Cellucci C. Rethinking knowledge. Metaphilosophy. 2015;46(2):213–34. https://www.jstor.org/stable/26602301

Cooper LN. Bound electron pairs in a degenerate Fermi gas. Phys Rev. 1956;104:1189. https://doi.org/10.1103/PhysRev.104.1189.

Einstein A. Quantum Theory of a Monoatomic Ideal Gas. A translation of Quantentheorie des einatomigen idealen Gases (Einstein, 1924). Translated by Amendola L. Retrievable at: www.equipes.lps.u-psud.fr/Montambaux/histoire-physique/Einstein-1924-I-anglais.pdf.

El-Khazri I. Disorder in Macbeth, or may we call it 'entropy'? Int J Human Cult Stud. 2015;2(3):255–61.

Fadeel B. Hide and seek: nanomaterial interactions with the immune system. Front Immunol. 2019;10:133. https://doi.org/10.3389/fimmu.2019.00133.

Feynman RP. There's plenty of room at the bottom. Eng Sci. 1960:22–36.

Forest V, Cottier M, Pourchez J. Electrostatic interactions favor the binding of positive nanoparticles on cells: a reductive theory. Nano Today. 2015;10(6):677–80. https://doi.org/10.1016/j.nantod.2015.07.002.

Grandjonc J, Risacher J, Maitron J. Caussidière Marc. 2009. Retrivable at: https://maitron.fr/spip.php?article28246.

Haken H. Synergetics. An introduction. Nonequilibrium phase transitions and self-Organization in Physics, chemistry and biology. Berlin Heidelberg: Springer-Verlag; 1978. https://doi.org/10.1007/978-3-642-96469-5.

Heurtebise JY. Vie et temps: dialectiques de l'ordre et du désordre. Sens public; 2008.

Hooke R. Micrographia: or some physiological descriptions of minute bodies made by magnifying glasses. In: With observations and inquiries thereupon. The Royal Society; 1665.

Hulla JE, Sahu SC, Hayes AW. Nanotechnology: history and future. Hum Exp Toxicol. 2015;34(12):1318–21. https://doi.org/10.1177/0960327115603588.

Jeevanandam J, Barhoum A, Chan YS, Dufresne A, Danquah MK. Review on nanoparticles and nanostructured materials: history, sources, toxicity and regulations. Beilstein Journal of Nanotechnology. 2018;9:1050–74. https://doi.org/10.3762/bjnano.9.98.

Joachim C, editor. Building and probing small for mechanics. Springer; 2020. https://doi.org/10.1007/978-3-030-56777-4.

Josephson BD. Possible new effects in superconductive tunnelling. Phys Lett. 1962;1(7):251–3. https://doi.org/10.1016/0031-9163(62)91369-0.

Josephson BD. The discovery of tunnelling supercurrents. Rev Mod Phys. 1974;46:251–4. https://doi.org/10.1103/RevModPhys.46.251.

Khan I, Saeed K, Khan I. Nanoparticles: properties, applications and toxicities. Arab J Chem. 2019;12(7):908–31. https://doi.org/10.1016/j.arabjc.2017.05.011.

Kundera M. L'ignoranza. Adelphi; 2003.

Leopardi G. I Canti. Laterza; 1917.

Lin C, Ke Y, Liu Y, Mertig M, Gu J, Yan H. Functional DNA nanotube arrays: bottom-up meets top-down. Angewandte Chemie International Edition English. 2007;46(32):6089–92. https://doi.org/10.1002/anie.200701767.

Lorenz EN. Deterministic nonperiodic flow. J Atmos Sci. 1963;20(2):130–41. https://doi.org/10.1175/1520-0469(1963)020<0130:DNF>2.0.CO;2.

Magni S, Miglio L, Milani M. Nanotechnology top-down: an introduction, from microelectronic to micromachining applications. Aracne; 2006.

Merriam-Webster. Crisis. Merriam-Webster.com Dictionary. Retrievable at https://www.merriam-webster.com/dictionary/crisis.

Malvaldi M, Leporini D. Capra e calcoli. L'eterna lotta tra gli algoritmi e il caos. Laterza; 2014.

Manoccio M, Esposito M, Passaseo A, Cuscunà M, Tasco V. Focused ion beam processing for 3D chiral photonics nanostructures. Micromachines. 2021;12(6) https://doi.org/10.3390/mi12010006.

Marcus M, Tuber S. Obsessive-compulsive disorder and uncertainty: struggling with a shadow of a doubt. Rowman & Littlefield; 2021.
Montagnini L. Harmonies of disorder. Norbert wiener: a mathematician-philosopher of our time. Springer International Publishing; 2017. https://doi.org/10.1007/978-3-319-50657-9.
Nicolis G, Prigogine I. Self-organization in non-equilibrium systems. From dissipative structures to order through fluctuations. Wiley; 1977.
Novotny L, Van Hulst N. Antennas for light. Nat Photonics. 2011;5(2):83–90. https://doi.org/10.1038/nphoton.2010.237.
Oakes. Chaos, Cosmos, and Logos. 2014. Retrievable at: http://faculty.winthrop.edu/oakesm/phil301/chaoscosmoslogos.pdf.
Preparata G. An introduction to a realistic quantum physics. Singapore: World Scientific; 2002.
Ruelle D. Where can one hope to profitably apply the ideas of chaos? Phys Today. 1994;47(7):24–30.
Ryu W-S. Molecular virology of human pathogenic viruses. Academic Press; 2017. https://doi.org/10.1016/C2013-0-15172-0.
Shakespeare W. Macbeth. Mondadori; 2004.
Snow CP. The two cultures and the scientific revolution. Cambridge University Press; 1959.
Svelto O. Principles of lasers. New York, NY: Springer; 2010. https://doi.org/10.1007/978-1-4419-1302-9.
Thibodeau PH, Boroditsky L. Metaphors we think with: the role of metaphor in reasoning. PLoS One. 2011;6(2):e16782. https://doi.org/10.1371/journal.pone.0016782.
Webb JH. Stimulated emission and Bose-Einstein statistics. Am J Phys. 1972;40(6):850–3. https://doi.org/10.1119/1.1986682.
Wigger H, Wohlleben W, Nowack B. Redefining environmental nanomaterial flows: consequences of the regulatory nanomaterial definition on the results of environmental exposure models. Environ Sci Nano. 2018;5:1372–85. https://doi.org/10.1039/C8EN00137E.

Part II

Plots and Connections of the Neverending Story

In this part, the mainframe of our investigation is described: the nanoworld where *Staphylococcus aureus* cells operate biodestruction on polyurethane, with electron microscopy as the main investigation tool. Although electron microscopy cannot cover all the relevant parameters of the system and the environmental features, it is a promising and effective tool: it can reach the magnification of interest to see membrane vesicles and their cargoes even without specific biochemical preparations.

The lifespan of nanoparticles generated by *Staphylococcus aureus* is illustrated in a typical environment, a tissue, where bacteria are active and give rise to inflammation and infection processes.

The delivery of xenobiotic material guided by the bacterial complex dynamics appears under a new light, opening new ways even to the possibility of delivering drugs to selected cells. A net of pathways associated with the bacterial secretion of proteins, cellular components, virulence factors, and a new kind of cargo is discussed raising issues in the nanomedicine field about the toxicological risks and the consequent necessary nanoparticles dosimetry protocols.

Staphylococcus aureus Scouts the Nanoworld: A Neverending Story

"The number of pages in this book is no more or less than infinite. None is the first page, none the last."
Jorge L. Borges

After the introduction into the nanoworld, the investigation will now focus on the lifespan of the nanoparticles generated by the biodegradative activity of *Staphylococcus aureus* cells operated on polyurethane dental prostheses, in a typical environment (a tissue) where bacteria are active and cause inflammation and infection. The process illustrated in Fig. 1, where a macroscopic object as bulk polyurethane is related to the micro- and the nano-domain, describes the possible dynamics that nanoparticles generated from the biodestruction process, as well as nanoparticles with totally different origin naturally present in biological structures, could undergo. In order to avoid a reductionist approach, a variety of technical and theoretical tools available from different areas of science is required.

Artificial or natural nanoparticles and membrane vesicles interact in a nonlinear way that open a series of paths in the dynamics of the nanoworld, some of them nontrivial and unexpected. In the nanosphere a series of processes and interactions happens not only between the two species of nanoparticles (nanoparticles and membrane vesicles) but also with ions, atoms, and molecules present in the medium, thus generating sometimes unpredictable dynamics and phenomena.

Considering the tissue with its bacterial hosts and the heterogeneous ensemble of nanoparticles as a whole, two observations are necessary; they are obvious but anyway have far-reaching implications on the behavior of the system itself and build the background for the toxicity issue that will be discussed in the forthcoming chapter. The first observation relies on the principle of mass conservation: the ensemble of the nanoparticles deriving from the biodestruction events must maintain the initial total mass, and while conserving it, nanoparticles can undergo accumulation processes, implying that all the particles or at least their fate should be traced. In mammalians these accumulation processes happen mainly in the liver and the spleen; studies on

M. Milani et al., *Bacterial Degradation of Organic and Inorganic Materials*,
https://doi.org/10.1007/978-3-031-26949-3_4

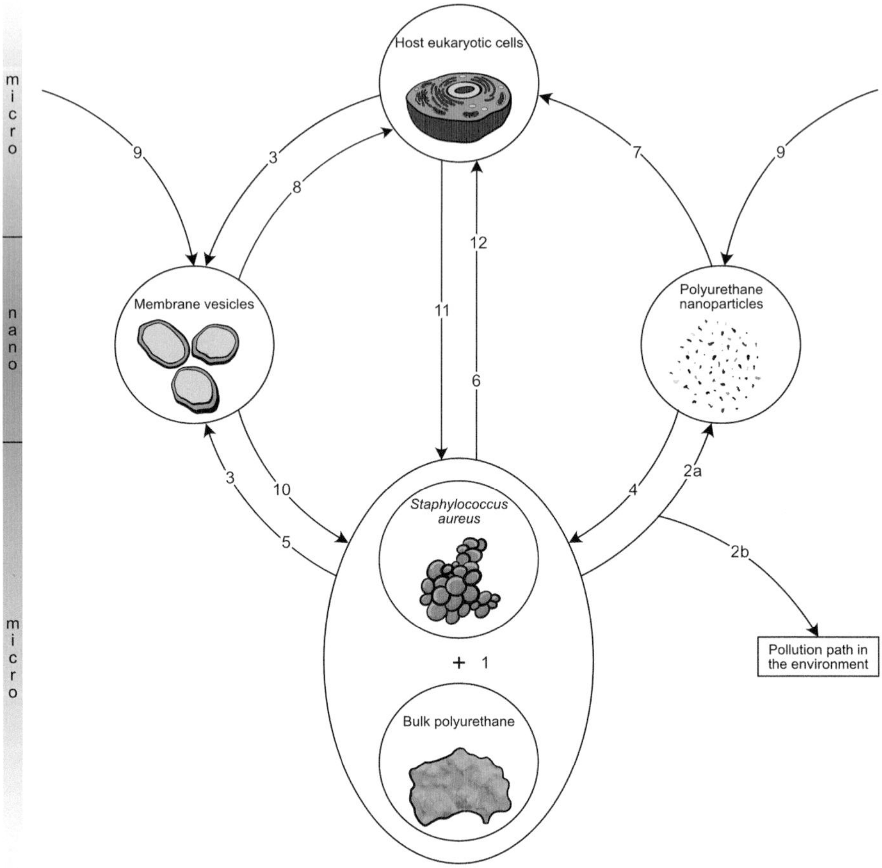

Fig. 1 Characters and processes of the neverending story. Main characters: *Staphylococcus aureus*; bulk polyurethane; polyurethane nanoparticles; membrane vesicles; host eukaryotic cells. Main processes: 1. Biofilm formation and biodestruction. 2a. Formation of polyurethane nanoparticles. 2b. Formation of debris. 3. Secretion of membrane vesicles. 4. Absorption of polyurethane nanoparticles from *Staphylococcus aureus*. 5. Secretion of membrane vesicles loaded with polyurethane nanoparticles. 6. Infection. 7. Host uptake of polyurethane nanoparticles. 8. Host uptake of bacterial membrane vesicles loaded or unloaded with polyurethane nanoparticles. 9. Formation of protein corona. 10. Absorption of loaded or unloaded membrane vesicles by *Staphylococcus aureus*. 11. Host response to staphylococcal infection. 12. Embedding of *Staphylococcus aureus* by host cells

chronic exposures and long term safety are required, especially because the evaluation of the nanoparticles toxicity sometimes leads to contradictory conclusions. A continuous uptake of nanoparticles in absence of discharge or expulsion processes from cells could lead to nanoparticle dynamics influenced by aggregation processes (with the nanoparticles moving from the nano- into the micro-domain); another route could lead to the secretion of nanoparticles from the system into the environment that therefore would act as a reservoir of nanoparticles (for instance they could be present in the deposits for drinking water derived from waste treatment) reinjecting them into the whole system represented in Fig. 1. As a consequence an attentive analysis of the toxicity levels is required because due to the constant increase of nanoparticles in cells, the trespassing of the LD50 can be expected also in the presence of constant low concentrations of nanoparticles. The second observation regards the interaction of the system with the environment and specifically the case of nanomedicine protocols based on nanoparticle insertion or dynamics of nanoparticle generation (Hussain et al. 2018; Curia et al. 2013; Schwarzenbach et al. 2010).

The scheme reported in Fig. 1 is built starting from the main characters of this story specifying the many processes they can undergo/provoke. The key points highlighted in Fig. 1 are the different possibilities for the researcher's choice as starting/ending points, the multicycle structure, and the "neverending" movement of the nanoparticles.

One of the possible starting points is the Staphylococcus aureus' biodestructive ability of bulk polyurethane with the generation of polyurethane nanoparticles, and their consequent interactions. These naturally generated polyurethane nanoparticles introduce different levels of complexity in the dynamics between nanoparticles and different biological components, cells in general, and their substructures (vesicles, organelles, cytoskeleton). Moreover the implications of the interactions between host cells are amplified by the pathological pathways of *Staphylococcus aureus* and its dynamics with immune host cells and evading strategies (Guerra et al. 2017; Otto 2014; Rasigade and Vandenesch 2014; Plata et al. 2009; Freeman-Cook and Freeman-Cook 2006; Foster 2005; Fedtke et al. 2004).

An overall vision of the system and its possible partitions in subsystems show the presence of a multiplicity of connecting paths that in certain cases appear as loops. The presence of loops points out the non-completeness of any approach based on the individuation of a unique starting point, the existence of feedback processes and the robust structural nonlinearity of the whole system, and, finally, the necessity of a clear identification of the approximations and assumptions. This is well represented by the theoretical description of a laser, a model for the collective dynamics of similar (identical) elementary components in interaction, which results can be often verified in experiments, and that can be considered equivalent to models proposed for autocatalytic chemical reactions (Haken 1983; Nicolis and Prigogine 1977; Sargent et al. 1970).

The abovementioned features are typical of different domains in science and humanities; several examples can be also found in literature where they look like mental scientific experiments (*Gedankenexperimente*) as in the case of *The Castle of Crossed Destinies*, a novel by the Italian writer Italo Calvino. The casual hand picking (operated by an individual or resulting from the cooperation of the members of a community) of cards from a tarot deck, each card endowed with its proper meaning and action rules, gives rise to a series of combinatorial situations that generate for the characters a catalogue (library) of possible paths to follow, involving a cascade of hypotheses, more or less realistic, and a multiplicity of outcomes (Calvino 1977). The narrator gives a teleological implant to the experiment by choosing both the starting point and the destination. In Calvino' story each character has values and meanings influenced by the place occupied in the temporal sequence that is arbitrary and dependent on the narrator (Manganelli 1973). The dynamics represented by Calvino and the competition/cooperation between random processes and mechanistic/deterministic ones find an experimental representation in the basic (elementary) behavior of a laser whose operation relies on a random, fundamental process (the spontaneous emission of photons) and on the further enhancement of the activity (optical amplification of photons by stimulated emission). The trials to visualize the tracks of the photons (resonant photons) in the laser cavity when moving from one excited atom to the other, in the simplest case of an atomic laser where two main kinds of loops are present (the absorption-spontaneous emission and the absorption-stimulated emission), can be useful illustrations of information associated to the basic features of the collective dynamics.

The similarity of the *Staphylococcus aureus'* nanoparticles neverending story and the world represented within the Calvino's novel has turning points in the interplay between randomness and determinism respectively ruled by the researcher and by the *corpus* of science related to the different characters. In this view the passage from a two-dimensional representation to a three-dimensional one in Fig. 1 can be useful and interesting, especially when the planar scheme becomes too complex and turns into an intricate mesh of connecting lines with a lot of intersections and knots.

As visible in Fig. 1, the presence of loops of different complexity and organization involves an increasing number of characters and processes that are strongly coupled. This makes clear the role of nonlinearity, feedback, and global possibilities of reversible processes: for instance nanoparticles (polyurethane nanoparticles or membrane vesicles) can move back and forth between bacterial and eukaryotic cells, or between bacterial cells and the environment, or again between eukaryotic cells and the environment in a sort of *perpetuum mobile*. This can be considered a typical issue in synergetics and can be extended to include collective phenomena and cooperation among not necessarily identical subsystems. This cooperation may lead to the formation of spatial, temporal, or functional structures. In this way a direct link in the interpretation and the dynamics (connection and distinction of cause and effect) between a picture (a static two-dimensional object) and time is established. Being the system an open one, as a result of rather unspecific changes of the environment, it acquires structures; in other words, the structures evolving in the system are not prescribed in a specific manner neither by the system itself nor by the outside, but derive from self-organization. Hence the formation of a structure can be interpreted as the emergence of new properties of the system associated to one or more order parameters (Haken 1991).

As already mentioned, the scheme in Fig. 1 represents a fairly complex open system where each character (*Staphylococcus aureus*, bulk polyurethane, polyurethane nanoparticles, membrane vesicles, host eukaryotic cells) can have different dictated and dictating facets, thus implying a greater complexity of the system. Considering the nanoparticles, they can be present with or without a protein corona and can be generated by processes of considerable importance as chemical-physical procedures or biosynthesis (engineered nanoparticles *tout court*, engineered nanoparticles derived by bacteria, engineered nanoparticles derived by *fungi*, engineered nanoparticles derived by viruses) (Prasad et al. 2018; Bittner et al. 2013; Nanda and Saravanan 2009). Also a reflection about the problem of the aggregation of nanoparticles is required: the process would result in the net reduction of the number of nanoparticles in the system in favour of the appearance of microparticles, which interaction with cells could be very different. Another element of complexity is introduced by the embedding of one or more polyurethane nanoparticles into membrane vesicles. As already mentioned membrane vesicles, due to their size, can be considered as nanoparticles with their own dynamics about which current knowledge is unaware of possible differences when in the presence of internalized nanoparticles (e.g. polyurethane nanoparticles).

Therefore, the real world and its complex dynamics are a combination (cooperation/competition) of all possible paths. This recalls the theme of the combinatory *ars* which is however made more complex by the possibility of the presence of both hypercycles, characterized by competition between cycles, usage of small selective advantage, increased information capacity and selection against parasitic branches and short circuits (Smith and Harper 1995; Eigen and Schuster 1978), and repetition, even *ad libitum*, that pushes the number of solutions to infinity. This fact leads the discussed problem into the domain of the halting problem and the issue of the real meaning of paths: in reality the trajectories are not lines (one-dimensional), but strips (strings) with extensions in space. In this way the number of possibilities can even increase, but the opposite can also be true. A further element to be considered is the main characteristic of the system, that of being open, an element related to the competition between computer treatments and analytical ones (Turing 1936, 1937, 2009).

> *Computers, when it comes to closed environments*
> *and with precise rules like a game, are unbeatable.*
> *Garri K. Kasparov.* (Authors' translation)

When aiming to the description of the behavior of closed systems, more properly defined compact sets (i.e., closed and limited), the sentence of Garri K. Kasparov brings to mind one of the fundamental elements characterizing quantum mechanics and in particular Schrödinger's equation, which admit discrete eigenvalues (with proper eigenstates) when the wave function is set in a Hilbert space compactly supported. When the system is non-closed and non-limited the eigenvalues are continuous (improper eigenstate), clearly reflecting the presence and role of the observer (in Kasparov's example represented by the chess player as opposed to the computer) that during the observation deforms the system (D'Alessandro 2019; Davydov 1976; Caldirola 1966).

The above considerations about loops, feedbacks, and crossing paths lead to the realization that the *Staphylococcus aureus* story is a "neverending" story with a labyrinthic conformation: a structure with infinite possibilities that implies the existence of a third type of time that overcomes the contradiction between the usual dichotomic perception and experience of cyclic and linear time, as it will be discussed in the last chapter of this book (Zaccaria Ruggiu 2006; Marramao 1992; Prigogine 1983).

The term labyrinth dates back to one of the most compelling Greek myths, the story of the Minotaur, half man and half bull, trapped in an unescapable maze. The labyrinth represents a structure built in such a way that it is difficult for those who enter to find the exit (or exits) (Halpern 2018). As an architecture of ambiguity and complexity that allows and/or prevents access promoting disorientation, the theme of the labyrinth is present in a lot of classical texts, for instance in the works of Ludovico Ariosto, Jorge L. Borges, Italo Calvino, Umberto Eco, and Michael Ende. Sometimes it happens that there is no single center; there might be no privileged connections, but broken, reversible, and interrupted lines, which intersect, meet, then diverge so much that, instead of divergence, it is more appropriate to speak of loops, highlighting the crucial role of chance.

Being necessary to frame the problem of the bacterial biodestruction in the proper environment, at this point it is suitable to mention the importance of the representation of Fig. 1 and its relations with the labyrinth. This step is fundamental to determine the right questions and to face the possibility of reaching interpretable results. This representation of the processes allows to understand the innovative content of the *Staphylococcus aureus* neverending story associated to the variety of types of connection between the characters, each of which loses the role of main actor.

More in detail historically the term labyrinth has geometrically different structures with important properties. Three kinds of labyrinth exist (Fig. 2):

- The classical or unicursal labyrinth (labyrinth of Knossos) where the entrance and exit correspond and there is only a route, so that all "points" are connected to a unique center.
- The manneristic labyrinth or Irrweg, where different routes are proposed. All routes but one are misdirecting, only one way leads to the exit that can or cannot correspond with the entryway. If unrolled, an Irrweg resembles an arborescent structure, with only one right way, where several mistakes can be corrected going backwards. No center can be identified and connections can be classified as right or wrong;
- The web labyrinth in which each point of a three-dimensional web can be connected to the other points through different ways. The aim is the exploration of the space, and the concept of wrong path does not exist.

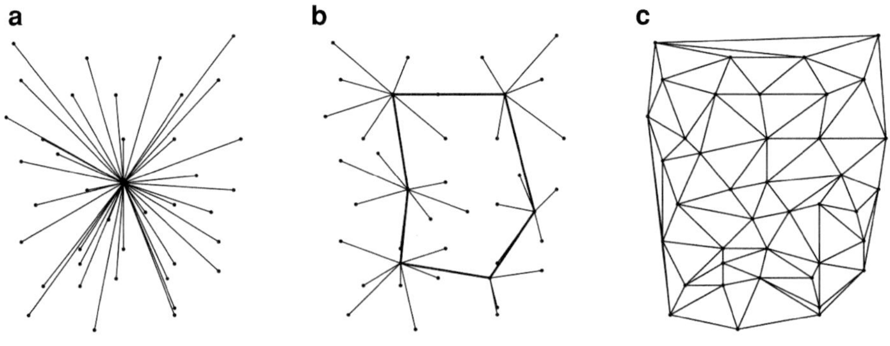

Fig. 2 Types of labyrinth generated by the same set of points (nodes). **a**) Centralized. **b**) Decentralized. **c**) Web-like

The first two types of labyrinth, the unicursal and the Irrweg, have both an interior and an exterior. The web labyrinth has neither interior nor exterior and the network model justifies and encourages the contradiction: all nodes can be connected, and loop processes can also occur.

Several scholars, including Eco, considered the labyrinth as a metaphor for men's attempt to trace and reproduce the complexity of the universe—one way represented by the unicursal labyrinth with only one possible path, the other by the multicursal version with many different roads, and the last most complex one represented by the rhizome with infinite possible paths (Eco 2014).

The labyrinth, with a relevant role also in science, manifests itself for instance in nature (Fig. 3) and is also present in a technique for the representation of multiple interactions (Feynman diagrams) fundamental in theoretical physics. Using space-time diagrams R. P. Feynman, who was a visual thinker, found that he could easily represent the interactions between electrons. In the drawings, the signs to the past seemed as logical as those to the future and from Feynman' standpoint the absence of causality had no impact on particle interactions since nothing established that the cause should always precede the effect (Halpern 2018). Interesting considerations on temporal reversibility in universes, the intervention on the arrow connecting causes and effects and the implication in real life can be found in John D. Barrow's *The Book of Universes* (Barrow 2012). In the same period while Feynman was developing the theory of the sum on the paths, the Argentine writer Borges published "The Garden of Forking Paths," a story focused on the theme of time intended as a labyrinth, where a sudden and unexpected turn reveals the volubility of the temporal flow, opening the possibility to multiple alternative processes (Halpern 2018; Borges 1999).

The connections between microscopic physical processes, their representations in diagrams, and their transformation into images are useful as a premise for the creation

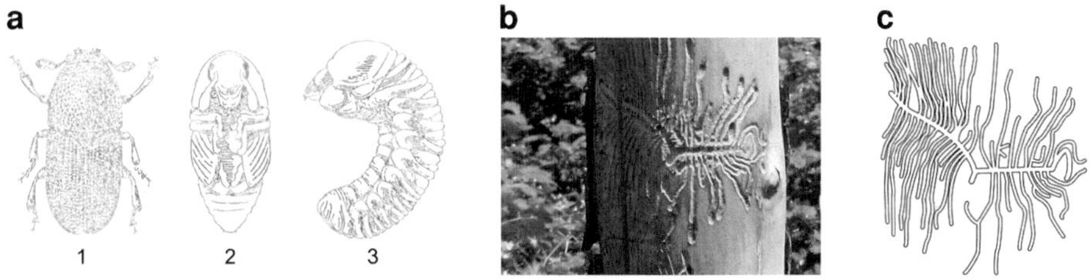

Fig. 3 From the tree to the labyrinth. Example of a natural labyrinthic structure made by *Scolytus rugulosus* (a) on the trunk of a *Prunus cerasus* (b). *Scolytus rugulosus* is a phytophagous that usually attacks debilitated orchard trees by drilling holes in the outer bark and starts creating 2-inch-long tunnels used to feed and reproduce. Adults usually dig clean tunnels that run parallel with the grain of the wood; at the sides of the tunnels at very short intervals female fellows gnaw little pockets in which they deposit about 80 eggs. The larvae hatch from the eggs and start excavating little longitudinal galleries that widen with the increasing size of the growing larva; the galleries progress sinuously and cross themselves in the final part (c). After about 20 days larvae reach maturity and dig a cell in the sapwood where they pupate. After 7–10 days adults exit directly through the bark (Chittenden 1898)

and reading of electron microscopy images. Considered as tools for representing and communicating results, images can have different origins; nevertheless they have common features including ambiguities, drawbacks, and space-time relations.

The labyrinthine structure is far from periodicity and linearity. The basic idea is that it is not written anywhere that something should move along a circular, straight, or curved path. A net is a system of lines that pervades a three-dimensional physical space where only the neighboring nodes are usually considered as correlated in short-range interactions. The example of biochemical and physical systems leads to the possibility of long-range interactions which enrich the phenomenological picture by introducing new phase transitions and new order parameters. These new dynamics can be characterized by a significant reduction of both the travel times and the lengths of correlation, allowing to pass from submicroscopic to macroscopic levels of operation as illustrated by several interesting physical examples (Del Giudice et al. 1986; Del Giudice et al. 1985; Fröhlich and Kremer 1983; Davydov 1981).

Actually, all the components of a system interact in all possible ways, in a mixture of choices. The advantages of the labyrinthine scheme are made evident by the analysis of the related properties:

(a) The story can start from anywhere.
(b) Each point or node can be connected to all the others.
(c) There is no single center.
(d) The story begins at one point and can follow different paths which however can intersect.
(e) The presence of multiple cycles or loops (which imply the lengthening of the connection times between the protagonists) and even of loops without output (infinite travel time).
(f) The infinite travel time could cause the labyrinthine structure to lose the informative content. In the previous chapter it was described how order can be associated with information (even if disorder does not automatically coincide with disinformation, misinformation, or error). The infinite travel time (which can be the time needed to pass from cause to effect) can be used in processes such as that of misdirection (i.e., disinformation or other typical techniques such as decoying). Disinformation has to be seen as inaccurate information that can be originated by incorrect and misleading statements, omissions, or conflict of interest, all behaviors that can exist without strict relations with the results of the investigations, the data.
(g) The links between the nodes can be fast tracks, as in the unicursal labyrinth with no way out other than the way in (the only way to exit the labyrinth is to go back, thus the only possible path is a unique line with no nodes on a surface), or slow and intricate as in a Irrweg maze due to the amount of the possible wrong ways (there can be more than one exit and every possibility does not include going back).
(h) There can be a lot of possible pathways made of lines crossed by numerous nodes that extend on a surface; this kind of maze is composed of half lines connected in a three-dimensional space that build a net labyrinth.

(i) The path can be deterministic but it can also show irregular oscillations, limiting the predictability of the future (chaotic or quasi-chaotic behavior).
(j) Reversibility and exchange between event (object)/news (track). All this has a close link with the cause-and-effect association and therefore with information, and superficially looks like a complex and exotic process, not belonging to daily life. The issue is a simplified analysis of information and its variants: disinformation, misinformation, and error which at first glance can be considered perfectly superimposable synonyms.
(k) "Perhaps one of the most important characteristics of the rhizome is that it always has multiple entryways" (Deleuze and Guattari 1987).
(l) In the net labyrinth every point can connect with any other node and the succession of connections can proceed unlimited. All the departing paths are potentially connected in a network of relationships that do not presuppose the uniqueness of the path, but its multiplicity. In other words, the rhizome is infinitely extendable. Moreover, among an unspecified range of alternative choices, even the wrong ones (at the same time opening the discussion about the definition and meaning of "error/mistake") produce solutions and together contribute to complicate the problem (Ronco 2011).

The Castle of Crossed Destinies, based on the ability of the reader to visualize the path described by the narrator, is both a literary and a visual *summa* of all the above properties (Calvino 1977).

News, facts, and data are the basis of both information and scientific activity. The possibility of wonder from information to disinformation or misinformation in the reading of schemes, maps, or figures requires a critical analysis. The discussion is more complicated when taking into account the complexity and possible reversibility of the cause-effect link.

> *The danger that I fear more than any other is the quantity of news and the impossibility of distinguishing information from disinformation: the flow is so high, unstoppable, that it prevents a plausible opinion from consolidating itself, from being supported by valid reasons.*
> *Salvatore Parlagreco.* (Authors' translation)

Normally news should derive from facts, i.e., come after the facts, but due to a temporal reversibility between events and news, the everyday experience shows how sometimes news surpass facts, in the sense that they can create facts actually preparing the way for them to happen. This inevitably leads to a state of ambiguity in which the news feeds the event that was triggered by the news itself. The ambiguity between facts and news becomes really elevated since sometimes when referring to facts it is better to think of the news, because news turn into facts and the only facts that matter are the news (Parlagreco 1990).

All the considerations made so far with the help of humanities, history, and literature, about information acquisition, interpretation, and transmission, are useful for the scientific research. In particular in the field of images, especially electron microscopy ones, even though the analysis is extremely attentive there are a lot of issues to take into account. The key point of image analysis is to search the truth or reality, but which is the role of images? Are they to be thought of as facts or news?

Recurring once more to literature, worth of mention is Luigi Pirandello's *The License*, in which a false news classifies the protagonist as a jinx, causing his transformation from a normal man to the one who brings bad luck. In this case the news is capable of transforming reality; in turn the facts generated by a man socially known as a jinx modify reality in a different way from those generated by a man considered a normal member of the society (Pirandello 2016).

The circularity between events, facts, and news affects the reading of the labyrinthine structure of the neverending story, actually playing a pivotal role in the journey towards knowledge (Fig. 4).

The circular relationship between news and events leads once more to talk about the role of misinformation that consists in providing a mass of information difficult to verify, and in orchestrating media campaigns in order to confuse true and false facts (Ceruso 2018). The mechanisms for building disinformation pass also through misdirection where the misleading activity is configured as a *post-factum* conduct, linked to a main event and aimed at protecting the real authors of the fact itself. This consolidated disinformation technique based on the skillful mixing of true, false, and plausible affirmations, (Ceruso 2018) results in decoying as a tool to transform news into facts which relevance is established by the appearance of the offense of misdirection punishable by the law.

The mixture of true, false, and plausible can be found in several fields, from the scientific one to the social and commercial areas. Disinformation thus constitutes a training field to power the ability of avoiding, detecting, and preventing such behavior in the field of image reading and interpreting.

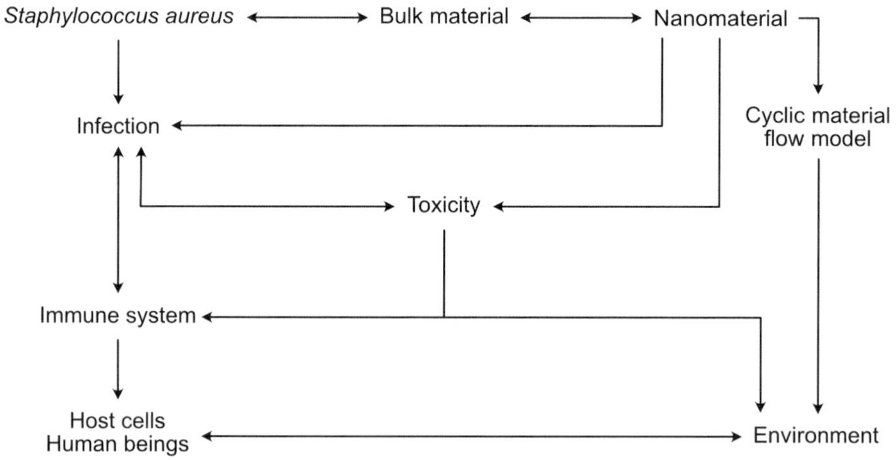

Fig. 4 Collocation of the neverending story in the real world. The processes illustrated in Fig. 1 are presented in the left part of this scheme. The circularity of the labyrinthine structure present in the *Staphylococcus aureus* neverending story undergoes further complications due to the infection mechanisms that enhance toxicity, and the interactions with the environment that promote the continuous exchange of nanoparticles also of external production

Several examples can be linked to disinformation, for instance the Stamina case where "Davide Vannoni, a psychologist turned medical entrepreneur" in a combined operation supported by a pharmaceutical/cosmeceutical company, in 2012 "has polarized Italian society with a bid to get his special brand of stem cell therapy authorized" (Abbott 2013; Arcovio 2013). The Stamina scam had a lot of follow-up due to the action of media, public figures, and reports from scientists (who afterwards in some cases changed their mind).

Another example of misinformation is given by the massive advertising with medical and journalistic support of the use of thermal water containing Radon for inhalations and baths. Self-explaining is the comparison between what is enthusiastically advertised by some thermal sites and what is reported in "Radon and Health," a public document of the World Health Organization retrievable at https://www.who.int/news-room/fact-sheets/detail/radon-and-health. Remaining in the radioactivity field, it is normal to find in the soil elements of Earth's crust, including some radionuclides in different concentrations, in commonly commercialized mineral waters. In Italy in the 1950s and 1960s the presence of radioactivity in bottled water was claimed to be beneficial to health, and until the 1980s it was normal to find indications of radioactivity on the labels of mineral water bottles. Only after the Chernobyl nuclear accident (1986), for marketing reasons, it was decided to avoid any indication related to radioactivity on the water labels. The disclosure of peculiar properties of water is the topic of a work by Henrik J. Ibsen *An Enemy of the People*, a play written more than a century ago in which in order to not collapse the economy and social structure of the spa town the truth cannot be told (which in the just mentioned case coincides with the chemical-physical analysis of water) (Ibsen 1999).

As a last example of disinformation, several papers can be considered inaccurate and misleading for the protection of public health (Hardell 2017; Starkey 2016). In a paper on Good Laboratory Practices there are evident contradictions about radiofrequency safety, since the authors on one hand claim that "a number of expert committees have concluded that there is no evidence for such risks as long as exposures are at or below levels that do not allow tissue heating," and on the other hand they also state that "electromagnetic fields research regarding possible health effects … does not adhere to specific guidelines and there are no standardized protocols for either the biological or for the exposure part of the studies" (Mattsson et al. 2021).

All the aforementioned cases are the peculiar exemplification of the importance of the process of data reading and interpreting, associated to the well-established procedure of assigning (sometimes arbitrarily) different weights to different data, also in relation with sets or subsets of metadata. Data reading and interpreting is the main activity implicit in looking at a micrograph. This element together with the consideration of the *Staphylococcus aureus* neverending story as a labyrinth of the third type has strategic implications in both the design of experiments and in formulating the right questions useful for understanding the active processes, so that the posed questions should not allow multiple or ambiguous answers, a fundamental issue in science.

As seen in the previous chapter, the main characteristics of nanoparticles raise important issues about their toxicity, bioaccumulation, and nanoparticle-induced pathologies. One of the most effective biological generators of nanoparticles are bacterial and fungal biofilms. Electron microscopy images are of great help in both bacterial characterization (especially FIB/SEM images) and in the investigation of the interaction of bacterial processes with different materials. In the current analysis images give detailed information about the characteristics of the supporting material and the interaction of *Staphylococcus aureus* with polyurethane. In Didenko et al., electron microscopy images show the steps of the biofilm formation, from the bacterial adhesion to the polyurethane surface after 24 h, to the formation of microcolonies after 48 h. Images testify the damages visible on the plastic surface after 30 days, and the consequent presence of micro- and nano-particles around the bacterial cell and of nanoparticles within membrane vesicles in the bacterial cells (Didenko et al. 2012).

Nanoparticles internalized within *Staphylococcus aureus* cells, when released by the bacterium inside a host tissue, in turn interact also with host eukaryotic cells. Although a large literature is present, it is still not completely known how nanoparticles interact either with the bacteria or with eukaryotic cells, but surely the dynamics of nanoparticles are governed by electromagnetic interactions. The electromagnetic forces are often overlooked due to theoretical and experimental difficulties, so that, though the acknowledgment that they can have a critical bias in the understanding of several processes, also in Fig. 1 electromagnetic forces are not explicitly mentioned.

One of the fundamental consequences of electromagnetic interactions is represented by the formation of the protein corona. When nanoparticles are in a biological fluid, a protein cover envelops the nanomaterial forming the protein corona. Physicochemical properties of the nanoparticles, which can be positively or negatively charged as well as neutral, the nature of the environment, and the duration of exposure can affect the composition and maturation of the protein corona, attracting or repelling biomolecules dispersed in the environment with different time scales. The presence of the corona cannot be confirmed by electron microscopy (and anyway this is a very complex issue that is still largely debatable), but it must be kept in mind that this is another element that modifies the nanoparticle charge and its approach to the bacterium. The protein corona affects the uptake of nanoparticles since alterations of fundamental electromagnetic properties modify what the cell sees (and consequently what is recorded on a micrograph and what can be seen by the researcher). This is an efficient concept that reorients research techniques shifting the point of view from the one of the researcher to the one of the biological character under investigation (bacterial or host eukaryotic cell). The uptake of nanoparticles by cells is at least a two-step process: the binding of the nanoparticle to the cell membrane and the subsequent internalization. The protein corona generates an additional level of dynamical complexity, a new interface, so that what the cell, either bacterial or eukaryotic, sees when interacting with a protein barrier matters more than the material characteristics of the nanoparticle. Hence the protein cover, shielding the nanoparticle and possibly acting as a camouflage, hiding the actual nature of the cargo, will determine the electrostatic and electrodynamic binding and subsequent internalization of xenobiotic nanomaterials within bacterial or eukaryotic cells (Forest et al. 2015; Walczyk et al. 2010; Lynch and Davey 2012; Lynch et al 2009). On the other hand, due to similar charge or different size and stability, proteins covering nanoparticles could not be affine to the bacterial uptake structures, thus inhibiting the internalization. The bacterial membrane, whose thickness is about 3–4 nm and whose resting membrane potential ranges between 30 and 70 mV, is characterized by a very high electric field up to 10^9 V/m; therefore the closeness of polyurethane nanoparticles (with or without the protein corona) to the bacterial cell can induce electrical dipoles and consequently determine the nanoparticle movement, approach, repulsion, and aggregation or adhesion in a complex nonlinear way (Curia et al. 2017).

Electron microscopy is a good tool to investigate the dynamics of nanoparticles. As visible in Curia et al. and Milani et al., from electron microscopy images it is evident that polyurethane nanoparticles in the proximity of bacterial cells or within membrane vesicles do not necessarily aggregate. Therefore nanoparticles, according to the attractive or repulsive interaction forces between them, can either remain free or group together. The possible aggregation of nanoparticles strongly affects the material internalization dynamics, moving the objects of observation from the nano- to the micro-domain. The analysis of electron microscopy images points out that only the nanoparticles (in the range of 2–10 nm) of polyurethane are actually internalized within the bacteria and that nanoparticles can be surrounded by membrane vesicles (of diameter ranging from 20 to 50 nm), thus giving hints about the uptake process (Fig. 5) (Curia et al. 2017; Milani et al. 2016).

The polyurethane nanoparticles' pathway from outside the cell to the interior of the bacterium can be influenced by many factors: the protein corona, colloidal forces, and dynamic biophysicochemical relations that can all be seen as complex combinations of elementary electromagnetic interactions. Two major features to be kept into account are the time dependence of all processes and the cellular and intracellular presence of the endogenous electromagnetic fields. There is a lot of literature about cell components, arrangement, movement, and dynamics but all the papers apart from some exceptions neglect the fundamental role of electromagnetism inside, astride, and outside the cell, giving credit to it only on the basis of electric charges and their Coulombian interactions, and on the basis of some interplay between electric rigid/permanent dipoles, more or less the basic "electrostatics" (Foster and Schwan 2019). In literature there are several reviews that present theories and experiments on how cells can generate, detect, and use electromagnetic fields in a wide range of frequencies, and of their possible mediation of cellular interactions. Several examples can be made, for instance the microbiology technique of electroporation or electropermeabilization in which an electrical field of the order of 10^6 V/m between two electrodes at a distance of 1–2 mm with a potential difference of about 1–1.5 kV is applied to cells in order to increase the permeability of the cell membrane. This allows the introduction into the cell of chemicals, drugs, or nucleic acids. Another interesting example is represented by the technique based on electrical shock for activation of *Xenopus* egg fertilization. Again, according to different states of the samples, it was measured via a cooled photomultiplier that electromagnetic waves in the range of 10^{14} Hz (optical radiation) can be emitted by biological samples (Angioletti-Uberti 2017; Mayburov 2015; Bersani 2012; Lynch and Davey 2012; Cifra et al. 2011; Musumeci et al. 2003; Popp and Zhang 2000; Dimitrov 1995; Fantes and Brooks 1993; Ho and Popp 1993; Fröhlich 1988; Colli et al. 1955).

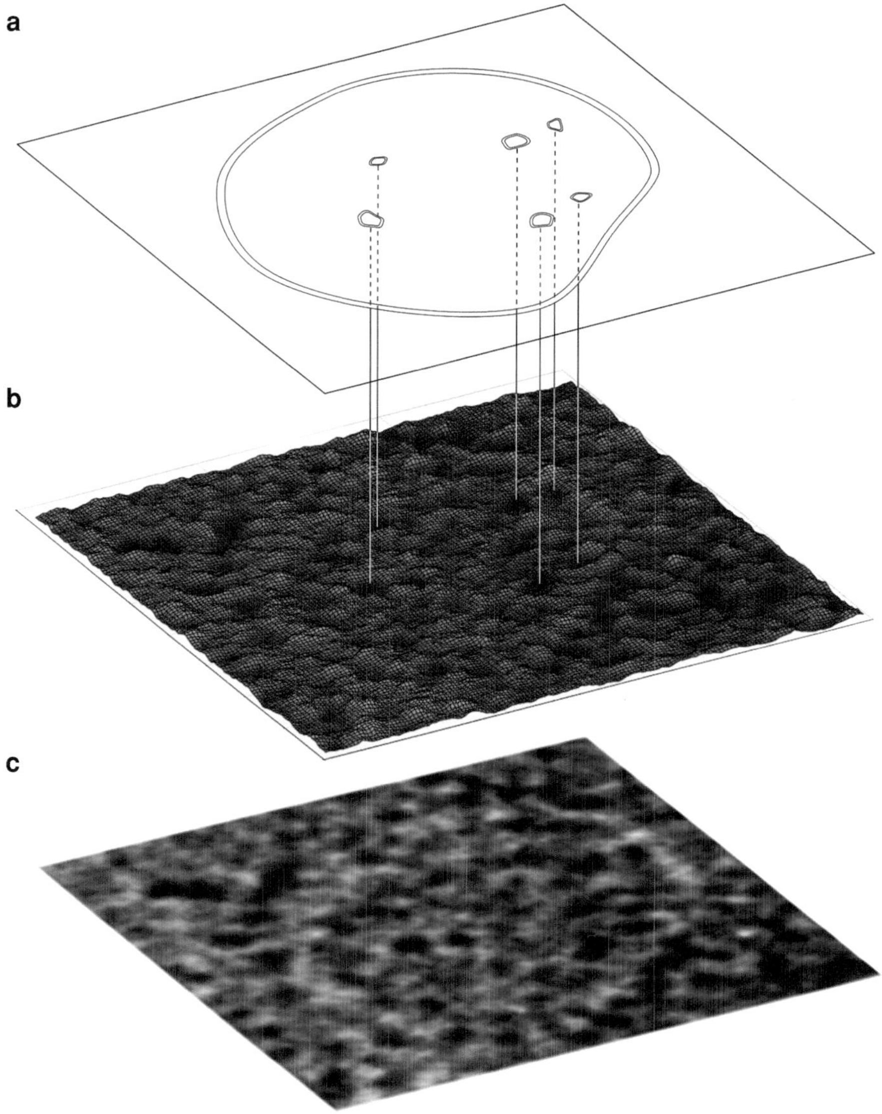

Fig. 5 Elaboration of a portion of a micrograph (Fig. 1a in Erega et al. 2017) of a membrane vesicle loaded with polyurethane nanoparticles within a *Staphylococcus aureus* cell. (**a**) Scheme of the polyurethane nanoparticles within the membrane vesicle for orientation purposes. (**b**) Plot of the membrane vesicle' surface: the lower grayscale values correspond to the denser portions of the sample such as membrane and nanoparticles. (**c**) TEM bright-field image of the membrane vesicle: electron-dense polyurethane nanoparticles (2–3 nm) appear darker than the surrounding components. The membrane vesicle's diameter is about 30 nm

As already explained, nanosized particles, detached from the polyurethane bulk surface as a consequence of the *Staphylococcus aureus* biodegradative activity, can be internalized into the bacterial cells. The absorption of polyurethane nanoparticles into *Staphylococcus aureus* occurs through endocytosis. The experiments give useful contributions in the understanding of how this process works in bacteria and can help to comprehend the role of the bacterial cytoskeleton, which is not only pivotal in the upkeep of the cellular structure, but in the regulation of the vesicular trafficking as well. The same experiments shed light also on the reverse process: the secretion of membrane vesicles loaded with nanoparticles. It is useful to remind that the uptake of nanoparticles by living cells is thought of as strongly size- and shape-dependent, with an optimal particle radius of ~ 25–30 nm at which the cellular uptake reaches a maximum. It is remarkable that the presence of polyurethane nanoparticles within *Staphylococcus aureus* does not compromise the bacterial viability; ultrastructural data prove that the internalization of nanoparticles does not prevent cell fission. Electron microscopy images show that membrane vesicles, loaded with one or more polyurethane nanoparticles, are arranged along a set of linear paths that connect the whole cell space. This pattern suggests the existence of bacterial internal structures with the function of highways or trails, with an important role of support for both the cell and the vesicular trafficking. Electron microscopy, revealing the vesicles' spatial distribution and providing a visual proof of the presence of privileged paths, suggests the presence of bacterial cytoskeletal structures (Curia et al. 2017; Milani et al. 2016; Curia et al. 2014; Didenko et al. 2012; Zhang et al. 2009; Ferrari 2008). The vesicular trafficking within bacterial cells is highly coordinated and complex. Membrane vesicles moving along microtubules can stop their movement and switch direction; moreover the switching between cytoskeletal structures is also possible. The electromagnetic features of this highly specialized intracellular transport system appear to transcend computational possibilities, reminding Roger Penrose's research about conscience and computability of mind processes strongly connected to cytoskeletal structures (Hameroff and Penrose 1996). There is a general lack of knowledge when it comes to bacterial strategy of vesicular transport; electron microscopy images of *Staphylococcus aureus* can help filling the gap elucidating steps of the vesicles' transport by the analysis of their spatial disposition with polyurethane nanoparticles that act as built-in markers (Curia et al. 2017; Erega et al. 2017; Milani et al. 2016; Curia et al. 2015). A classification of different types of vesicles and their functions can be done. The classical endocytic pathway starts with early endosomes (the vesicular compartment that first receives the incoming cargo) that, moving along microtubules towards the center of the cell, mature and convert into late endosomes. They are the linking point between the recycling pathway (late endosomes retrograde at early endosomes) and the degradation cycle (late endosomes undergo a further maturation into lysosomes) (Huotari and Helenius 2011). Analyzing electron microscopy images, it is not possible to make the same classification for the bacterial membrane vesicles, but it is likely that if a vesicle loaded with polyurethane nanoparticles is found outside a bacterial cell, that very vesicle was previously inside the bacterium. In this case polyurethane nanoparticles act as markers, testifying that vesicles which have absorbed the polymer were once within the bacterium.

The presented results from *in vitro* experiments suggest some hypotheses regarding the *in vivo* dynamics of nanoparticle-host cell interactions with implications in the toxicological field. Polyurethane nanoparticles surrounded by membrane vesicles within bacteria represent a new way through which xenobiotic nanomaterials may gain access to human organs with the risk of bioaccumulation. The secretion of membrane vesicles could appear as an expensive process in terms of energy for bacteria, but vesicles derived from bacterial cells can bring advantages to microorganisms. As seen in Chap. 1, membrane vesicles could be responsible for the *Staphylococcus aureus*' high resistance against antibiotics acting as a decoy for antibodies, making them actually non-effective against the whole microorganism; moreover membrane vesicles have a key role in the bacterial pathogenesis that includes the transfer of several types of cargo (proteins, phospholipids, signaling molecules, virulence factors, nucleic acids, and genetic information) without the activation of the immune system, the involvement in cell signaling and the formation of biofilm, cytotoxicity, and the invasion of eukaryotic host cells with consequent cause of infection and disease (Cao and Lin 2021; Askarian et al. 2018).

In an *in vivo* hypothetical infection occurring in a human being where *Staphylococcus aureus* cells have absorbed polyurethane nanoparticles, bacterial cells, able to evade the immune system, become a new vector for nanoparticles, providing them a new way of access to body tissues. Moreover nanoparticles could be either discharged by the bacterium or released as a consequence of the bacterial death, and thus exit the bacterial cell with or without the membrane vesicle. In addition electron microscopy images show the presence of nanoparticles stuck on the bacterial cell wall: once the bacterium attacks the host cell, these nanoparticles could promptly access the host cell also in absence of a vesicle (Curia et al. 2017). The destiny of bacterial-generated nanoparticles raises issues in the toxicological field about the bioaccumulation levels of polyurethane in human tissues and organs, toxicological risks related to the nanosize of the material, and the possible pathologies provoked. In this way the neverending story of *Staphylococcus aureus* continues as an infinite cycle and can represent the story of any other nanomaterial. The ability of nanoparticles to move inside the body sets a great threat when associated to the potential hazard of the nanomaterial, the particles' size, and bioaccumulation as pathogenicity factor. Moreover interactions between nanoparticles and biosystems broaden the spectrum of other possible mechanisms of absorption such as inhalation, ingestion, dermal penetration, or intravenous perfusion. Further investigations are required also for the nanoparticle-based drug delivery systems that seem to offer advantages over current therapeutic agents that commonly display a longer circulation time, lower toxicity, specific targeted release, and greater bioavailability. The biological consequences of the interactions between nanoparticles and biosystems need to be evaluated taking into account the context, so it becomes clear that the focus on the crucial problems must be directed to both the biomedical and the nanosafety fields (Fig. 6) (Raza et al. 2019; Hussain et al. 2018; Forest et al. 2015).

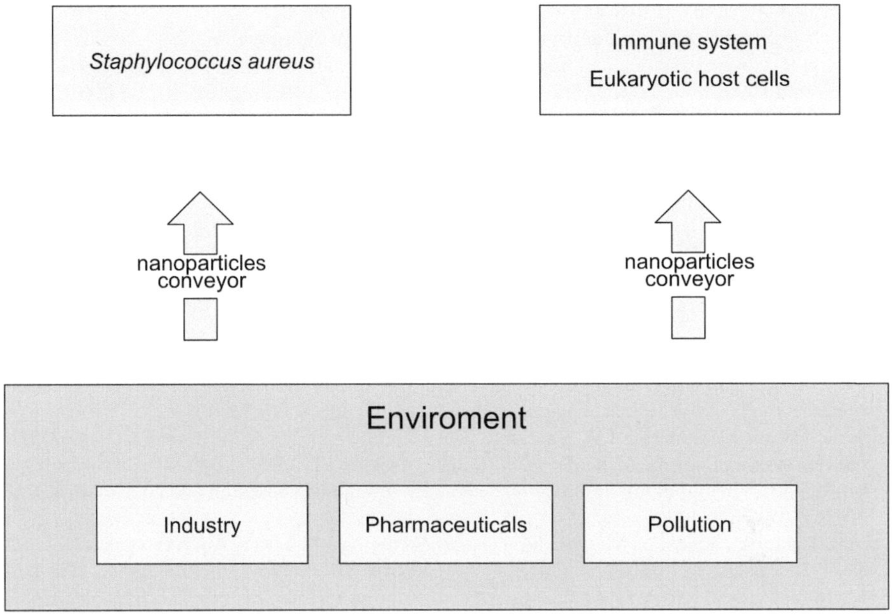

Fig. 6 Areas involved by the neverending story. The scheme shows the different fields interested by the biological consequences of the interactions between nanoparticles and biosystems that regard not only individuals but affect the whole environment with the circularization of nanoparticles of different origin

References

Abbott A. Italian stem-cell trial based on flawed data. Nature. 2013; https://doi.org/10.1038/nature.2013.13329.

Angioletti-Uberti S. Theory, simulations and the design of functionalized nanoparticles for biomedical applications: a Soft Matter Perspective. NPJ Computut Mat. 2017;3(48):1. https://doi.org/10.1038/s41524-017-0050-y.

Arcovio V. Dietro Stamina una multinazionale della cosmetica. La Stampa: I soldi e gli interessi internazionali sul metodo Vannoni; 2013.

Askarian F, Lapek JD Jr, Dongre M, Tsai CM, Kumaraswamy M, Kousha A, Valderrama JA, Ludviksen JA, Cavanagh JP, Uchiyama S, Mollnes TE, Gonzalez DJ, Wai SN, Nizet V, Johannessen M. *Staphylococcus aureus* membrane-derived vesicles promote bacterial virulence and confer protective immunity in murine infection models. Front Microbiol. 2018;9:262. https://doi.org/10.3389/fmicb.2018.00262.

Barrow JD. The book of universes. Vintage; 2012.

Bersani F, editor. Electricity and magnetism in biology and medicine. Springer Science & Business Media; 2012.

Bittner AM, Alonso JM, Górzny MŁ, Wege C. Nanoscale science and technology with plant viruses and bacteriophages. In: Mateu M, editor. Structure and Physics of Viruses, Subcellular Biochemistry, vol. 68. Dordrecht: Springer; 2013. https://doi.org/10.1007/978-94-007-6552-8_22.

Borges JL. Collected fictions. Penguin Books; 1999.

Caldirola P. Istituzioni di Fisica teorica. Editrice Viscontea; 1966.
Calvino I. The castle of crossed destinies. Martin Secker & Warburg Ltd; 1977.
Cao Y, Lin H. Characterization and function of membrane vesicles in gram-positive bacteria. Appl Microbiol Biotechnol. 2021;105:1795–801. https://doi.org/10.1007/s00253-021-11140-1.
Ceruso V, La mafia nera. I depistaggi tra eversione neofascista e cosa nostra: storia di un'Italia oscura. Newton Compton; 2018.
Chittenden FH, The Fruit-tree Bark-beetle, *Scolytus rugulosus* Ratz., United States Department of Agriculture, Division of Entomology, 1898; Circular 29.
Cifra M, Fields JZ, Farhadi A. Electromagnetic cellular interactions. Prog Biophys Mol Biol. 2011;105(3):223–46. https://doi.org/10.1016/j.pbiomolbio.2010.07.003.
Colli L, Facchini U, Guidotti G, Lonati RD, Orsenigo M, Sommariva O. Further measurements on the bioluminescence of the seedlings. Experientia. 1955;11:479–81. https://doi.org/10.1007/BF02166829.
Curia R, Erega A, Milani M. Study of the formation, Secretion and Trafficking of Membrane Vesicles Loaded with Polyurethane Nanoparticles in *Staphylococcus aureus* via Electron Microscopy. HSOA J Nanotechnol Nanomed Nanobiotechnol. 2015;2:005.
Curia R, Milani M, Didenko LV, Avtandilov GA, Shevlyagina NV, Smirnova TA. Beyond the biodestruction of polyurethane: *S. aureus* uptake of nanoparticles is a challenge for toxicology. In: Méndez-Vilas A, editor. Microscopy: advances in scientific research and education, vol. 1; 2014. p. 16–23.
Curia R, Milani M, Didenko LV, Avtandilov GA, Shevlyagina NV, Tatti F. Analysis of the Bacterial Vesicles' Enhanced Toxicological Threat Via Electron Microscopy. In: Talukdar K, Bhushan M, Malipatil AS, editors. Advancing medicine through nanotechnology and nanomechanics applications. IGI Global; 2017. p. 19–44. https://doi.org/10.4018/978-1-5225-1043-7.ch002.
Curia R, Milani M, Didenko LV, Shevlyagina NV. Electron microscopy broadens the horizons of toxicology: the role of nanoparticles vehiculated by bacteria. Curr Topics Toxicol. 2013:9.
D'Alessandro J, Kasparov: Il computer vincerà sempre e non è un male, La Repubblica, 16/05/2019.
Davydov AS. Biology and quantum mechanics. Pergamon Press; 1981.
Davydov AS. Quantum mechanics. Pergamon Press; 1976.
Del Giudice E, Doglia S, Milani M, Vitiello G. A quantum field theoretical approach to the collective behaviour of biological systems. Nuclear Physics B. 1985;251:375–400. https://doi.org/10.1016/0550-3213(85)90267-6.
Del Giudice E, Doglia S, Milani M, Vitiello G. Electromagnetic field and spontaneous symmetry breaking in biological matter. Nuclear Physics B. 1986;275(2):185–99. https://doi.org/10.1016/0550-3213(86)90595-X.
Deleuze G, Guattari F. A thousand plateaus. Capitalism and Schizophrenia: University of Minnesota Press; 1987.
Didenko LV, Avtandilov GA, Shevlyagina NV, Smirnova TA, Lebedenko IY, Tatti F, Savoia C, Evans G, Milani M. Biodestruction of polyurethane by *Staphylococcus aureus* (an investigation by SEM, TEM and FIB). In: Méndez-Vilas A, editor. Current microscopy contributions to advances in science and technology. Formatex Research Center Publishing; 2012. p. 323–34.
Dimitrov DS. Electroporation and electrofusion of membranes. In: Lipowsky R, Sackmann E, editors. Handbook of biological physics, vol. 1. North-Holland; 1995. p. 851–901.
Eco U. From the tree to the labyrinth: historical studies on the sign and interpretation. Harvard University Press; 2014.
Eigen M, Schuster P. The Hypercycle. Naturwissenschaften. 1978;65:7–41. https://doi.org/10.1007/BF00420631.
Erega A, Curia R, Milani M. Interactions between nanoparticles and bacterial structures as seen through software enhanced analysis of electron microscopy images. In: Méndez-Vilas A, editor. Microscopy and imaging science: practical approaches to applied research and education. Formatex Research Center Publishing; 2017. p. 587–94.
Fantes P, Brooks R. The cell cycle: a practical approach. IRL Press at Oxford University Press; 1993.
Fedtke I, Götz F, Peschel A. Bacterial evasion of innate host defenses – the *Staphylococcus aureus* lesson. Int J Med Microbiol. 2004;294:189–94. https://doi.org/10.1016/j.ijmm.2004.06.016.

Ferrari M. Beyond drug delivery. Nat Nanotechnol. 2008;3:131–2. https://doi.org/10.1038/nnano.2008.46.
Forest V, Cottier M, Pourchez J. Electrostatic interactions favor the binding of positive nanoparticles on cells: a reductive theory. Nano Today. 2015;10(6):677–80. https://doi.org/10.1016/j.nantod.2015.07.002.
Foster KR, Schwan HP. Dielectric properties of tissues. In: Polk C, Postow E, editors. CRC handbook of biological effects of electromagnetic fields. CRC Press; 2019. p. 27–96.
Foster TJ. Immune evasion by *staphylococci*. Nat Rev Microbiol. 2005;3(12):948–58. https://doi.org/10.1038/nrmicro1289.
Freeman-Cook L, Freeman-Cook KD. *Staphylococcus aureus* infections. Infobase publishing; 2006.
Fröhlich H, editor. Biological coherence and response to external stimuli. Berlin Heidelberg: Springer-Verlag; 1988. https://doi.org/10.1007/978-3-642-73309-3.
Fröhlich H, Kremer F, editors. Coherent excitations in biological systems. Berlin Heidelberg: Springer-Verlag; 1983.
Guerra FE, Borgogna TR, Patel DM, Sward EW, Voyich JM. Epic immune battles of history: neutrophils vs. *Staphylococcus aureus*. Frontiers in Cellular and Infection Microbiology. 2017;7:286. https://doi.org/10.3389/fcimb.2017.00286.
Haken H. Synergetic computers and cognition. A top-down approach to neural nets. Berlin Heidelberg: Springer-Verlag; 1991. https://doi.org/10.1007/978-3-662-22450-2.
Haken H. Synergetics. An introduction. Berlin, Heidelberg: Springer; 1983. https://doi.org/10.1007/978-3-642-88338-5.
Halpern P. The quantum labyrinth. In: How Richard Feynman and John Wheeler Revolutionized Time and Reality. Basic Books; 2018.
Hameroff S, Penrose R. Orchestrated reduction of quantum coherence in brain microtubules: a model for consciousness. Math Comput Simul. 1996;40(3–4):453–80. https://doi.org/10.1016/0378-4754(96)80476-9.
Hardell L. World Health Organization, radiofrequency radiation and health-a hard nut to crack. Int J Oncol. 2017;51(2):405–13. https://doi.org/10.3892/ijo.2017.4046.
Ho M-W, Popp FA, Biological organization, coherence, and light emission from living organisms , in: Stein WD, VarelaFJ (Eds.), Thinking about biology, CRC Press, 1993:183–214.
Huotari J, Helenius A. Endosome maturation. EMBO J. 2011;30:3481–500. https://doi.org/10.1038/emboj.2011.286.
Hussain S, Joo J, Kang J, Kim B, Braun GB, She Z-G, Kim D, Mann AP, Mölder T, Teesalu T, Carnazza S, Guglielmino S, Sailor MJ, Ruoslahti E. Antibiotic-loaded nanoparticles targeted to the site of infection enhance antibacterial efficacy. Nature Biomedical Engineering. 2018;2:95–103. https://doi.org/10.1038/s41551-017-0187-5.
Ibsen HJ. An enemy of the people. Dover Publications; 1999.
Lynch I, Salvati A, Dawson KA. What does the cell see? Nat Nanotechnol. 2009;4:546–7. https://doi.org/10.1038/nnano.2009.248.
Lynch PT, Davey MR, editors. Electrical manipulation of cells. Springer Science & Business Media; 2012.
Manganelli G. Postfazione. In: Calvino I, editor. Il castello dei destini incrociati. Einaudi; 1973.
Marramao G. Kairós. In: Apologia del tempo debito. Laterza; 1992.
Mattsson MO, Simkó M, Foster KR. 5G new radio requires the best possible risk assessment studies: perspective and recommended guidelines. Frontiers in Communications and Networks. 2021;2 https://doi.org/10.3389/frcmn.2021.724772.
Mayburov SN. Photon emission and quantum signalling in biological systems. EPJ Web of Conferences. 2015;95:03024. https://doi.org/10.1051/epjconf/20159503024.
Milani M, Didenko LV, Avtandilov GA, Curia R, Erega A, Shevlyagina NV. Electron microscopy documents the microorganisms' biodestructive action on polyurethane and the production, internalization and vesicular trafficking of nanoparticles. Br J Appl Sci Technol. 2016;12(3):1–19. https://doi.org/10.9734/BJAST/2016/20290.

Musumeci F, Brizhik LS, Ho M-W. Energy and information transfer in biological systems. how physics could enrich biological understanding. In: Proceedings of the International Workshop, Acireale, Catania, Italy, 18–22 September, 2002. World Scientific Publishing; 2003.

Nanda A, Saravanan M. Biosynthesis of silver nanoparticles from *Staphylococcus aureus* and its antimicrobial activity against MRSA and MRSE. Nanomedicine. 2009;5(4):452–6. https://doi.org/10.1016/j.nano.2009.01.012.

Nicolis G, Prigogine I. Self-organization in non-equilibrium systems. In: From dissipative structures to order through fluctuations. Wiley; 1977.

Otto M. *Staphylococcus aureus* toxins. Curr Opin Microbiol. 2014;17:32–7. https://doi.org/10.1016/j.mib.2013.11.004.

Parlagreco S. Il mistero del corvo. SugarCo: Il caso Di Pisa; 1990.

Pirandello L. Henry IV: followed by the license. Italica Press; 2016.

Plata K, Rosato AE, Węgrzyn G. *Staphylococcus aureus* as an infectious agent: overview of biochemistry and molecular genetics of its pathogenicity. Acta Biochim Pol. 2009;56(4):597–612. https://doi.org/10.18388/abp.2009_2491.

Popp FA, Zhang J. Mechanism of interaction between electromagnetic fields and living organisms. Science in China Series C Life Sciences. 2000;43(5):507–18.

Prasad R, Kumar V, Kumar M, Wang S. Fungal nanobionics: principles and applications. Singapore: Springer; 2018. https://doi.org/10.1007/978-981-10-8666-3.

Prigogine I, Time, life and entropy, International Seminar on the Living State 1983, Printed and published for Professor Mishra RK.

Rasigade JP, Vandenesch F. *Staphylococcus aureus*: a pathogen with still unresolved issues. Infect Genet Evol. 2014;21:510–4. https://doi.org/10.1016/j.meegid.2013.08.018.

Raza A, Sime FB, Cabot PJ, Maqbool F, Roberts JA, Falconer JR. Solid nanoparticles for oral antimicrobial drug delivery: a review. Drug Discov Today. 2019;24(3):858–66. https://doi.org/10.1016/j.drudis.2019.01.004.

Ronco M, Semantica ed estetica del labirinto , Parol on line, 2011. Retrievable at: http://www.parol.it/articles/Manuela-Ronco_Semantica-ed-estetica-del-labirinto.pdf.

Sargent M, Scully MO, Lamb WE. Buildup of laser oscillations from quantum noise. Appl Opt. 1970;9(11):2423–7. https://doi.org/10.1364/AO.9.002423.

Schwarzenbach RP, Egli T, Hofstetter TB, von Gunten U, Wehrli B. Global water pollution and human health. Annu Rev Environ Resour. 2010;35:109–36. https://doi.org/10.1146/annurev-environ-100809-125342.

Smith MJ, Harper DG. Animal signals: models and terminology. J Theor Biol. 1995;177(3):305–11. https://doi.org/10.1006/jtbi.1995.0248.

Starkey SJ. Inaccurate official assessment of radiofrequency safety by the advisory group on non-ionising radiation. Rev Environ Health. 2016;31(4):493–503. https://doi.org/10.1515/reveh-2016-0060.

Turing AM. Computing machinery and intelligence. In: Epstein R, Roberts G, Beber G, editors. Parsing the Turing test. Dordrecht: Springer; 2009. https://doi.org/10.1007/978-1-4020-6710-5_3.

Turing AM. On computable numbers, with an application to the Entscheidungsproblem. J Math. 1936;58:345–63.

Turing AM. On computable numbers, with an application to the Entscheidungsproblem; a correction. Royal Society; 1937.

Walczyk D, Bombelli FB, Monopoli MP, Lynch I, Dawson KA. What the cell sees in bionanoscience. J Am Chem Soc. 2010;132(16):5761–8. https://doi.org/10.1021/ja910675v.

Zaccaria Ruggiu A, Le forme del tempo. Aion, Chronos, Kairos , Il poligrafo, 2006.

Zhang S, Li J, Lykotrafitis G, Bao G, Suresh S. Size-dependent endocytosis of nanoparticles. Adv Mater. 2009;21(4):419–24. https://doi.org/10.1002/adma.200801393.

Nanoparticles and Toxicity

Often think of the rapidity with which things pass by and disappear,
both the things which are and the things which are produced.
For substance is like a river in a continual flow...
and there is hardly anything which stands still.
Marcus Aurelius Antoninus

Nanotoxicology investigates the potential toxic effects of nanomaterials and nanoparticles on the environment and human health. Nanotoxicology aims to quantitatively evaluate the danger associated to the use of nanomaterials, studying their production processes and the possible exposure routes. In this chapter only a short list of factors relevant to the issue is presented and will be discussed as follows in order to determine the recommendations for risk prevention. Numerous materials which are nontoxic in bulk exhibit intense toxicity as their size is reduced. Due to the increasing applications of nanotechnology in several fields (industry, medicine, life sciences, pharmaceuticals, electronics, and energy production) and the extensive use of nanomaterials, the investigation of the potential adverse effects due to nanoparticle exposure is a fundamental issue.

As seen in Chap. 3, thanks to peculiar characteristics as size, shape, and surface charge, that give them unprecedented advantages, nanoparticles are widely used. Special regard to the consequences on health and safety provoked by exposures to nanomaterials need to be investigated because they could result in undesirable immunological effects. With respect to their bulk form, at the nanosize materials, especially in biomedical applications, have a much higher surface/volume ratio that facilitates their inhalation, ingestion, and absorption, and hence allows nanoparticles to move inside the body towards non-target cells, tissues, and organs (heart, liver, kidneys, spleen, bone marrow, and nervous system) passing barriers, thus representing a major threat (Curia et al. 2017; Curia et al. 2013).

M. Milani et al., *Bacterial Degradation of Organic and Inorganic Materials*,
https://doi.org/10.1007/978-3-031-26949-3_5

Nanoparticles, often due to industrial processes, are also released into the environment (water, soil, air) causing ecological toxic effects that irremediably have an impact on the human and animal food chains (Walters et al. 2016; Buzea et al. 2007).

Toxicity of Nanoparticles

Characteristics and processes that enhance the toxicity of nanoparticles are various, such as particle size, surface area and charge, shape/structure, aggregation/dissolution, and surface coatings. The physical and chemical characteristics of nanoparticles, such as magnetic, optical, thermal, mechanical, and electrical properties, make them suitable for several applications. However, nanoparticles can deeply affect the human body and the environment. For instance, metal-based nanoparticles commonly used in biomedicine, may affect cells inducing mutations, DNA oxidation, reduced cell viability, morphology defects, apoptosis and necrosis, and decreased proliferation; in addition in some cases metal nanoparticles may persist in the organisms. Due to the enormous differences that exist between classes of nanoparticles, it is difficult to generalize the possible toxic effects of nanomaterials, and an in-depth analysis of the properties of each nanosized material is mandatory.

The size of a nanoparticle deeply affects its distribution and consequently its toxicity, and since the particles' reactivity is strictly dependent on their size, particles of the same material but with different dimensions can deposit in diverse places.

The kinetics of nanoparticles is affected by their morphology, being the shape a fundamental parameter in the nanoparticle uptake. Nanoparticles can have different forms (spheres, rods, cubes, pyramids) moreover, also the surface charge, somehow introducing a virtual shape not based on mass distribution, but instead related to the charge distribution, biases size measurements and determines the particles' dispersion, thus influencing the adsorption of ions and biomolecules.

The aggregation of nanoparticles is another issue deeply related with the material toxicity. This process implies a change in the size and surface/volume ratio of the nanomaterial that can cause unreliable dosimetry data since particles could accumulate in the body in the form of aggregates, therefore being no longer detectable as nanoparticles.

Nanoparticle dissolution determines possible toxic effects in the aquatic environment.

The characterization of the physical and chemical properties that influence the behavior and toxicity of nanomaterials is important since it can clarify how nanoparticles correlate with ecological or biological environments. When discharged in the environment nanomaterials might undergo modifications which could eventually increase or decrease their toxicity. Moreover, also external factors such as temperature, pH, salinity, or the presence of organic matter can affect the dynamics of nanoparticle. pH affects the nanoparticle surface charge and as a consequence also the nanomaterial's mobility and aggregation.

The morphological characterization of nanomaterials is possible via TEM, SEM, AFM, and STM. The elemental analysis is usually conducted using an EDX. Dynamic light scattering or electrophoretic light scattering spectroscopy are used to determine an effective size of nanoparticles in fluids, whereas X-ray powder diffraction is used for nanoparticles in the dry state; the zeta potential is useful to characterise the nanoparticles' features especially in biomedical applications.

Nanoparticles easily penetrate cells and other biological barriers of living organisms causing cell damage. There are several routes that allow nanomaterials to enter cells. Inhalation of airborne nanoparticles is the most common. Depending on the shape, size, and aggregation state, nanoparticles are deposited in the respiratory tract or in the lungs from which nanomaterials may enter the bloodstream and hence reach any body district. Depending on the class, inhaled nanoparticles can cause inflammation, granulomas, and pulmonary fibrosis that can lead to carcinogenic effects and systemic cardiovascular deficiency. The effects on the lungs are obvious and significant, so a protective action for people exposed to nanodust is required.

Nanoparticles can be ingested due to their abundant presence in food or due to an unintentional hand-to-mouth transfer of materials.

Another route of exposure is through dermal penetration. Nanoparticles find easy access to the body through wounds; besides, according to their size, shape, water solubility, and surface coating, nanoparticles are also able to enter the body through the intact skin and provoke the release of pro-inflammatory cytokines, cause oxidative stress and decreased viability.

The behavior and the interactions of nanoparticles inside the body are a big issue that could trigger stress reactions, weakening the body's defense and leading to inflammation processes.

One of the fundamental aspects to keep in mind in the toxicological field is the bioaccumulation of nanomaterials inside the human body; especially the interactions of non-degradable particles with the biological reactions within the body should be evaluated.

Depending on their composition and concentration, nanomaterials can have negative effects at different levels of cellular organization, such as increased oxidative stress, inflammatory cytokine production, and cell death.

The high chemical reactivity of nanomaterials is associated to an increased production of reactive oxygen species (ROS). This may lead to oxidative stress, i.e., an imbalance between the production of ROS and the cells' ability to reduce them. The damages provoked by oxidative stress include oxidative modification of proteins, generation of protein radicals, initiation of lipid peroxidation, modification of nucleic acids, modulation of gene expression, cell death, and genotoxic effects.

In general nanomaterials can affect the cellular viability, hindering cellular functions and leading to the cell death; one of the possible effects is represented by apoptosis as a consequence of mitochondrial damage and oxidative stress.

Nanomaterials have the ability to cause damages to the genetic material. The nanoparticle-induced genotoxicity can result in chromosomal fragmentation, DNA strand breakages, point mutations, oxidative DNA adducts, alterations in gene expression, hereditary diseases, promotion of mutagenesis, and carcinogenesis.

Biodegradation is the capacity of microorganisms, both bacteria and fungi, to attack bulk inert material to produce debris of different sizes, from micro- to nanoparticles. The microorganisms' biodegradative activity on polyurethane dental prostheses, described so far, highlights how the research and development of the materials used in biomedical applications needs to be investigated in-depth to obtain more biocompatible, durable, and resistant materials. Thanks to their versatility, excellent mechanical properties, and good biocompatibility, polyurethanes have generated a growing interest and are commonly used in medical applications such as general hospital tubing, stents, and orthopedic prostheses; moreover polymeric nanoparticles have shown potentiality for drug delivery and medical imaging (Bhavsar et al. 2023; Huang et al. 2018; Cooper and Guan 2016).

In light of the current knowledge about the bacterial pathogenicity and the biodegradative action on polymeric dental prostheses, the *Staphylococcus aureus* neverending story, described in the previous chapter, should be taken as a stimulating guide for further investigations on the materials' properties and duration, and on the generation and fate of polyurethane nanoparticles. Details of the neverending story should help researchers to focus their attention on the precise aspects of the problem, especially related to the nanotoxicological field which still lacks information about bioaccumulation levels, toxicological risks, exchanges among bacterial cells, host cells and reversible transfer from cells of one type to the other, and possible pathologies provoked by nanomaterials. The study of the nanomaterials' surface functionalization by different chemical groups that can either provoke or attenuate the immune responses of the body to nanomaterials is a critical element for their biomedical safety and efficacy. Polyurethane nanoparticles generated from microorganisms need to be better investigated since the characterization of nanomaterials produced by bacteria, differing from the engineered polyurethane nanoparticles for size, surface charge, and physicochemical properties, would be extremely useful in the understanding of the interaction processes between bulk polyurethane and bacteria, polyurethane nanoparticles and bacteria, and polyurethane nanoparticles and host eukaryotic cells, thus providing information also about the potential of polyurethane nanoparticles in medical therapies. The mechanisms of absorption and the interactions between nanoparticles and biosystems, together with the ability of nanosized materials to move inside cells, pass barriers, and hence potentially reach any part of the organism, need to be fully understood, since the potential hazards of both the nanomaterial and the bioaccumulation as pathogenicity factor are crucial problems for biomedicine and nanosafety (Curia et al. 2017; Curia et al. 2013).

All living bodies are constantly exposed to xenobiotic nanomaterials in one way or another. The wide use of nanomaterials in consumer products and the potential risks for both the environment and human health are increasing concern. The precise characterization of the nanoparticles' features and properties is of fundamental importance to understand their dynamics and biological effects. The task of nanotoxicology is surely challenging, since not only the properties of the different types of nanoparticles need to be analyzed in-depth, but also the properties of biological systems at the nanoscale that interact with nanoparticles are not completely clear, as for instance the exposure concentrations and bioaccumulation in tissues and the role of bacterial membrane vesicles as vehicles for nanomaterials.

Electron microscopy imaging is of great help in the investigation of nanotoxicity as a diagnostic tool; therefore it should accompany every *in vitro* experiment. Among the investigating instruments electron microscopy is one of the more direct because it operates at the dimension scale of interest, and, despite the ambiguities and uncertainties of the interpretation process as it will be discussed in the forthcoming chapter, it provides a straightforward visual information as an electron microscopy image in one of the forms more linkable to the observed object.

New methodologies able to assess the presence and reactivity of nanoparticles in commercial, environmental, and biological samples should be developed. Also for this topic the neverending story is relevant since *Staphylococcus aureus*, with its ubiquity and its role as modifier, transporter and injector of nanoparticles of different origin towards various targets (sometimes unexpected and unpredictable), well-represents the complexity of the problems of defining the nanoparticles' exposure limits. Therefore a regulatory framework to control the dosimetry and the risks associated to the intensive use of products containing nanoparticles should be assessed. To achieve any significant improvement in the nanotoxicological field, a joint multidisciplinary cooperation in several fields of medical sciences, nanotechnology, biology, biomedical engineering and physics should be established, and it is for this reason that this chapter has been inserted without any pretense to be a review of the field of nanotoxicology, but just to outline all the possible viable routes of such an extensive problem as the bacterial biodegradation of polymeric prostheses and the consequent implications.

(Bobyk, 2021; Akçan, 2020; Ganguly et al. 2018; Huang et al. 2018; Cooper and Guan 2016; Amedea, 2015; Powers et al. 2007; Nel et al. 2006; NIOSH n.d.).

References

Akçan R, Aydogan HC, Yildirim MŞ, Taştekin B, Sağlam N. Nanotoxicity: a challenge for future medicine. Turkish J Med Sci. 2020;50(4):1180–96. https://doi.org/10.3906/sag-1912-209.

Amedea B. Nanotoxicology of metal oxide nanoparticles. Mdpi Ag. 2015;5:934–75. https://doi.org/10.3390/met5020934.

Bhavsar P, Bhave M, Webb HK. Solving the plastic dilemma: the fungal and bacterial biodegradability of polyurethanes. World J Microbiol Biotechnol. 2023;39;122. https://doi.org/10.1007/s11274-023-03558-8.

Bobyk L, Tarantini A, Beal D, Veronesi G, Kieffer I, Motellier S, Valsami-Jones E, Lynch I, Jouneau P-H, Pernet-Gallay K, Aude-Garcia C, Sauvaigo S, Douki T, Rabilloud T, Carrière M. Toxicity and chemical transformation of silver nanoparticles in A549 lung cells: dose-rate-dependent genotoxic impact. Environ Sci Nano. 2021;8(3):806–21. https://doi.org/10.1039/D0EN00533A.

Buzea C, Pacheco II, Robbie K. Nanomaterials and nanoparticles: sources and toxicity. Biointerphases. 2007;2(4):MR17–71. https://doi.org/10.1116/1.2815690.

Cooper SL, Guan J, editors. Advances in polyurethane biomaterials. Woodhead Publishing; 2016. https://doi.org/10.1016/C2014-0-04143-3.

Curia R, Milani M, Didenko LV, Avtandilov GA, Shevlyagina NV, Tatti F. Analysis of the bacterial vesicles' enhanced toxicological threat via electron microscopy. In: Talukdar

K, Bhushan M, Malipatil AS, editors. Advancing medicine through nanotechnology and nanomechanics applications. Hershey: IGI Global; 2017. p. 19–43. https://doi.org/10.4018/978-1-5225-1043-7.ch002.

Curia R, Milani M, Didenko LV, Shevlyagina NV, "Electron microscopy broadens the horizons of toxicology: The role of nanoparticles vehiculated by bacteria", Current Topics in Toxicology, 2013;9:93–8.

Ganguly P, Breen A, Pillai SC. Toxicity of nanomaterials: exposure, pathways, assessment, and recent advances. ACS Biomater Sci Eng. 2018;4(7):2237–75. https://doi.org/10.1021/acsbiomaterials.8b00068.

Huang YJ, Hung KC, Hung HS, Hsu SH. Modulation of macrophage phenotype by biodegradable polyurethane nanoparticles: possible relation between macrophage polarization and immune response of nanoparticles. ACS Appl Mater Interfaces. 2018;10(23):19436–48. https://doi.org/10.1021/acsami.8b04718.

Nel A, Xia T, Madler L, Li N. Toxic potential of materials at the nanolevel. Science. 2006;311(5761):622–7. https://doi.org/10.1126/science.1114397.

NIOSH, The National Institute for Occupational Safety and Health. Retrievable at: https://www.cdc.gov/niosh/index.htm.

Powers KW, Palazuelos M, Moudgil BM, Roberts SM. Characterization of the size, shape, and state of dispersion of nanoparticles for toxicological studies. Nanotoxicology. 2007;1(1):42–51. https://doi.org/10.1080/17435390701314902.

Walters C, Pool E, Somerset V. Nanotoxicology: a review. Toxicol New Aspects Sci Conundrum. 2016; https://doi.org/10.5772/64754.

Part III

Voyage in the Interpretation of Images

Yes, we will question everything, everything once again.
And we won't rush ahead with seven-league boots,
but crawl at a snail's pace.
Berthold Brecht

This last chapter illustrates the starting points and the questions (that have the value and the role of answers themselves) deriving from the major themes of the book through a little journey in the world of images and their power, and aims to be a rational guide in a labyrinthic world abundantly polluted by "fast truths" facilitated by digital technology and its often over estimated authority.

Since the whole book and especially this chapter are a special thanks not only to the scientific skills but also to the culture and humanity of Dr. Didenko, a few verses of Alexander S. Pushkin which she quoted during one of our usual conversations after the daily laboratory activity will be the common linking thread.

Oh, how many wonderful discoveries
Enlightenment spirit prepares for us!
And Experience, the son of difficult errors,
And the Genius, the friend of paradoxes,
and the Case, the God is the inventor.[1]

The culture and humanity of Dr. Didenko appear to be reinforced by the following Latin quote written in the poster mentioned in the introduction.

Vita brevis, ars vera longa, occasio autem praeceps, experientia fallax, iudicium difficile.

This quote tells us a lot about the importance of culture as both a unicum and a unifying factor: features that are gone missing over the last few centuries but retrievable every now and then not only in bridges across different knowledge areas (science, art, humanities) but also across time.

[1] Pushkin AS, "Collected works in 10 volumes, 1959–1962", Vol. 2. Poems 1823–1836.

The origin of the phrase is a Greek quote by Hippocrates of Kos.

Ὁ βίος βραχύς,
ἡ δὲ τέχνη μακρή,
ὁ δὲ καιρὸς ὀξύς,
ἡ δὲ πεῖρα σφαλερή,
ἡ δὲ κρίσις χαλεπή.

It is commonly rendered into English as "Life is short, the art long, opportunity fleeting, experiment treacherous, judgement difficult," a translation related to the Latin adaptation of some *ad hoc* selected verses taken up by Lucius A. Seneca "*vitam brevem esse, longam artem*"[2] who with his cut strongly influenced and even distorted the deep meaning of the original sentence.

Notwithstanding the gap of twenty-two centuries, the interpretation of the Hippocrates' aphorism and the one of Pushkin's verses connect and even melt. The sets of words present in the two quotes not only overlap and partly coincide but taking into account the width of the meaning that can be attributed to each one of them due to the different languages (thus highlighting the important role of translation) and historical periods, they complement each another.

In his verses, Pushkin pointed out that human beings start their lives making *discoveries*; *education* and culture are the pillars of their growth and the premises for construing their own selves and surroundings. Life dotted with choices and *mistakes* will inspire doubts and will lead to the building of one's own burden of *experience* that, linking and mixing apparently unconnected notions, will give rise to knowledge.

You were not made to live as brutes,
but to follow virtue and knowledge.
Dante Alighieri

According to the chosen path, people will face situations, facts and sensations, at times odd, ambiguous (where chance conflicts with hazard) and even paradoxical, i.e. sometimes inconsistent, impossible, and bewildering. This attitude is a good viaticum to handle the many cross-linked themes of this chapter, a fascinating adventure where hazard, mistakes, knowledge, experience, truth (*doxa*), and paradoxes mix together, with the final aim of interpreting micrographs.

According to the interpretation of Hippocrates, experience can be enriching, but it can be also fraught with hazards. The presence of different Latin variants of uncertain origin underlines the relevant role of interpretation that relates to and influences the idea people have of life; in this frame, experience sometimes generates discoveries with a questionable reliability (*experientia fallax*), other times derives from or is associated with painful or difficult mistakes (*experimentum periculosum*), not only from the scientific standpoint but also from the human one.

It is not surreal that a scientist as Dr. Didenko acknowledged the importance of the thought of both a medical doctor considered "the father of the Western modern

[2] Seneca LA, "*De brevitate vitae*", Mondadori, 2016.

Medicine" and one of the greatest poets of the nineteenth century, considered the "Sun of the Russian poetry," acute observer of the world and lover of culture. Pushkin's art, able to describe with wittiness and irony the "things of life," is still nowadays of topical interest; after all people in science and in art are engaged in the same enterprise and share the same language.[3,4,5,6,7]

> What makes up the truth?
> Doubt. Truth cannot be found in our rationality if there is no doubt.
> Giacomo Leopardi

As Giacomo Leopardi,[8] contemporary to Pushkin and so similar yet so different from him, pointed out doubt (intended as accepting ambiguity as an essential aspect of knowledge) is the key for the pursuit of truth, the element without which looking behind at all the traces left on the path would not be a possibility. In this fundamental quest help is provided by past experiences: the attention snatches from something meaningless just moments before, suddenly a combination of loops and straight lines with a plausible cause–effect relation is created, and finally the chance strikes and the individual discovers new routes in the intricate web of connections in the labyrinth of knowledge. Images, sounds, and sensorial experiences in general acquire multiple interpretative significances. Although we stress again its crucial importance, we leave open the question about the fact that images can be "representative" or "typical," and we do not discuss whether the images (microscopy images in our case) due to poor design of the experiment and/or subconscious bias of the investigator give a misleading impression or not.

To conclude, the role of time, with its properties of periodicity and irreversibility, must not be neglected. On the one hand, these features must be implemented by the right/good time (*Kairos* in Greek mythology) related to synchronicity and coherence. On the other hand, nature is usually described by time-dependent (with few exceptions of subfields devoted to memory processes, duty cycles, hysteretic and nonequilibrium processes) fundamental equations, symmetrical with respect to time reversal just like space variables, but *Kairos* evidently covers a critical role in photography shooting and images reading, at the basis of scientific explorations.

---

[3] Rozentuller V, Talbott S, "From Two Cultures to One: On the Relation between Science and Art", Context 13, 2005:13–18.

[4] Sapov VV, "Pushkin is our contemporary", Herald of the Russian Academy of Sciences, 2008;78(6):520–521. https://doi.org/10.1134/S1019331608060063.

[5] Riezler K, "Physics and Reality. Lectures of Aristotle on Modern Physics at an International Congress of Science", Yale University Press, 1940.

[6] Boncinelli E, Giorello G, "L' incanto e il disinganno: Leopardi. Poeta, filosofo, scienziato", Guanda, 2016.

[7] Grammaticos PC, Diamantis A, "Useful known and unknown views of the father of modern medicine, Hippocrates and his teacher Democritus", Hellenic Journal of Nuclear Medicine, 2008;11(1):2–4.

[8] Massarenti A, "Italy24 book of the week: Giacomo Leopardi's "Zibaldone"", 2015, retrievable at: http://www.italy24.ilsole24ore.com/print/ABvkDCuC/0?refresh_ce=1

Images, Knowledge, and Doubt

Открытие—Discovery

The only true voyage of discovery ... would be not to visit strange lands but to possess other eyes.
Marcel Proust

During the weeks spent in the laboratory of Doctor Lyubov V. Didenko, it was usual for her and Professor Marziale Milani to conclude a working day chatting and sipping tea in the break room while the rest of the crew headed home. Discussions were not a mere summing up of the work and the results of the day, but free talks which often included personal reflections. This triggered a lot of correspondences and personal discussions on the role of science in society, and it was not uncommon to start from technicalities to eventually shift to general problems in life sciences, medical cases, or even philosophy and history, sometimes invading the domains of poetry.

The cooperation between Milan and Moscow's laboratories grew stronger through this Proustian "voyage of discovery" and strong was the feeling of the necessity of building a bridge between the two major branches of culture: science and humanities. Although often seen as two sometimes conflicting sides of a dichotomic culture (Snow 1959) which reflects also on society, the link between these two domains is not rare, as illustrated by, among others, Heraclitus of Ephesus, Titus Lucretius Carus, Dante Alighieri, Leonardo di ser Piero da Vinci, Giordano Bruno, Galileo Galilei, Giacomo Leopardi, Vasily V. Kandinsky, Bertolt Brecht, Primo Levi, and Isaac Asimov.

One day the discussion about electron microscopy images, their meaning, reliability and what one can really see in them dealt with the subject of image interpretation; surely at the time the complexity of the topic was overlooked, but that chat was an incentive to analyze in depth the process of image production

M. Milani et al., *Bacterial Degradation of Organic and Inorganic Materials*,
https://doi.org/10.1007/978-3-031-26949-3_6

and interpretation, with special attention not only to the education acquired in years of studies but also to one's personal origins and cultural background. That same day before leaving for the evening Dr. Didenko quoted the following Alexander S. Pushkin[1]'s verses, without offering any translation or comment about them:

О сколько нам открытий чудных
Готовят Просвещенья Дух
И Опыт, сын ошибок трудных,
И Гений, парадоксов друг,
И Случай, бог изобретатель.
А.С. Пушкин (1829)

А.С. Пушкин. Собрание сочинений в 10 томах. М.: ГИХЛ, 1959–1962. Том 2. Стихотворения 1823–1836

A.S. Pushkin. Collected works in 10 volumes, 1959–1962. Volume 2. Poems 1823–1836

It was a challenge for Professor Milani to understand these five verses, firstly due to the obstacle of the different language and secondly because the contextualization in the electron microscopy field was hard to find. Back to Italy, Internet provided a lot of translations (from Russian to English) all different from one another and finding out what Dr. Didenko meant became more and more intriguing. Furthermore, one of Pushkin's poetic rules was "not to spell everything out - this is the secret of arousing interest" (Nilsson 1987).

The team of Italian researchers pushed by those five verses with so different translations and as many interpretations, and led by the pleasure of the discovery, decided to apply the same steps necessary to translate a text in a foreign language to images, especially to electron microscopy ones, and soon the connections between text, translation and interpretation arose. After all, every type of communication, be it a text or an image, is based on translation. Images are considered the "primordial condition of the translation" because they can picture an object, which afterwards will be named by a word and described by a sign (Sini 1989).

As different translations make it hard to take in the author's message, so different interpretations of the same represented object can lead to fuzzy results and mistakes. One of the elements at the basis of this journey is the importance to start from the investigation of the acquiring process of images (micrographs, photographs, paintings, etc.), following with the analysis of the resulting figures to finally arrive to their interpretation. In order to better understand the features of this path, the thought went to the most familiar form of images: the photograph. It was

[1]Alexander Sergeyevich Pushkin (June 6, 1799 [O.S. 26 May]–February 10, 1837 [O.S. 29 January]) was a Russian writer, considered the founder of the modern Russian literature. On the day of his birthday, UNESCO (United Nations Educational, Scientific and Cultural Organization) established the Russian Language Day to celebrate multilingualism and cultural diversity. His works span from narrative poems and dramas to prose and fairy tales, but the author was familiar also with the scientific method and had a serious interest in natural sciences (as supported by the presence in his library of books of astronomy and medicine). In Pushkin's works poetic inspiration and scientific insight are indistinguishable; the poet in search of the perfect word architecture and musicality is the symbiont of the scientist in search of nature's mysteries.

John F. W. Herschel, astronomer and experimental photographer, that in 1839 coined the term photography (from Greek φῶς photo, light and γραφή graphia, writing) which means "writing with light," and metaphorically "drawing with nature," since at the time the only light source with enough power was the sun. This way of thinking photographs was so stimulating that in 1844 William H. F. Talbot, scientist and photography pioneer, published a book titled *The Pencil of Nature* (Rogers 2002).

From the beginning of their life, people are immersed in a world of images and, as time passes by, get assailed by them always more impressively and massively. Daily image bombardments of every kind (advertisings, magazines, television, mass media) have transformed and elevated the role of images which apparently have become more communicative than words, a feature that often can be found in scientific papers motivated in many different ways. But are we really able to dig under layers of easily catchable interpretations to understand the actual meaning of the images we are looking at? Or do images force people, also those animated by a sincere love for knowledge and beauty, to deny their faculty of thought and expression to surrender to the bore of receiving ready-made hardly biased and influencing notions? (Huizinga 1935)

As Richard L. Gregory, psychologist and founder of the Department of Machine Intelligence and Perception at the University of Edinburgh (UK), said: "images are useless shadows of objects until their significance is read" (Gregory 2001). This is a provocative but precious statement: on one hand it makes us move out of the comfort zone of our brains, and on the other hand it focuses on the word "shadow." What this term represents has multiple meanings in scientific, figurative and literal domains. The short story *Avatar* by Théophile P. J. Gautier is a good example to appreciate all the diverse senses intended every time the word "shadow" was used, as it is the story by Jack London *The Flash and the Shadow* or the work titled *Shadows* by the art historian Ernst H. J. Gombrich where many different artistic meanings have been attributed to the term (Gombrich 1995; London 1903; Gautier 1857). Shadows in Plato's *Allegory of the Cave* become useful whenever a meaning is attributed to them, but they lack a univocal correspondence with objects' part of the real world, the first difference being the number of spatial dimensions of shadows as compared to those proper of real objects (Plato, ~ 375 BC). Using the terminology of semiotics, a field often neglected by science, the shadow becomes a sign which means something, but in the case of the cave the sign is linked to uncertainty and limited knowledge (Eco 2016).

> *Itaque duo sunt genera imaginum. Aliæ enim sunt similes rebus extrinsecis secundum totum vel per integrum... Aliæ vero sunt similes rebus extrinsecis secundum partes, sed non secundum totum.*
>
> *And so there are two kinds of images. One consists of images similar to external objects in their entirety or wholeness... The other one consists of images that reflect parts of real external objects but not actual existing things in their totality.*
>
> *Giordano Bruno*

So how can an image switch from shadowy to understandable? To understand what an image means, we need to approach the world of image interpretation starting from comprehending what an image is and how an image represents, and only after these two steps we can try to learn how to read its meaning.

Since the time of Greek philosophers, images were thought of as a true copy of reality, but the nature of this relationship was a source of controversy: according to Plato an image was an imperfect copy of the material reality, whereas Aristotle thought that an image could imitate nature (mimesis) (Ogden 2010; Carbone 2009). The Latin term for image "*imago, -inis*" is not linked only to the visual and optical sphere, in fact it finds translations in a lot of fields from vision, dream, apparition, and even echo, with an evident recall to magic, witchcraft, and sorcery.

> It is a sad reality … that most of us can only see what, due to previous knowledge, imagine finding.
>
> *Neil McGregor*

In his essay *De Anima* Aristotle introduced a new cognitive function called *phantasia*, the faculty that imagination uses to elaborate sensations awakened both from the five senses and thought. Two contrasting representational modes spring from the imaginative element: one essentially mimetic, rooted in the representational power of artistic forms, the other abstract and figurative, dependent on or linked to a non-existent image grasped through the soul's imagination. The mimetic mode is thoroughly conventional and seems to pose few problems of interpretation; the figurative one requires a total different and distinctive form of attention (Kemp 1977).

During the action of looking at an image the eyes gather a series of information that the brain processes. The system reaches the most plausible global interpretation integrating cues from the retinal input, but what we perceive is actually much more complex than a simple topographical representation of things; the brain has a key role registering details and features of what is depicted, assigning them different weights, characteristics that could be essential during actions like thinking and elaborating memories.

One of the topical issues of this book is the extraction of reliable information (knowledge) from a picture. In the case of electron microscopy images of the biodegradative action of *Staphylococcus aureus* on polyurethane surfaces, observers can see the presence of nanoparticles; the same picture can also be seen as a set of data that observers arrange in different ways on the basis of a "paradigm" (an ensemble of values, beliefs, theories, techniques). Data in science (often supplemented by a certain amount of metadata), being the result of a series of operations, choices, refinements, and procedures made by the researcher, are not independent from the interactions and perspective of the person studying a specific phenomenon. Therefore, observations are forcibly influenced by the researcher's interests. Visual processing is one of the most complex tasks achieved by humans since visual skills include not only recognition, but also the guidance of actions and the creation and manipulation of visual images (Bokulich and Parker 2021; Leonelli 2019; Schurz 2013; Nanay 1997).

If we look at a dazzling altogether colorless object, it makes a strong lasting impression, and its after-vision is accompanied by an appearance of color.
Johann W. von Goethe

The German poet and scientist Johann W. von Goethe had full "knowledge of the phenomena of subjective colors and described various modes of producing them." Von Goethe encouraged creativity, experience, and genius; he thought that the observation of nature could lead to the understanding of the self, and, for this reason, the knowledge of nature could not be forced. According to von Goethe, darkness "had so much to do as light with the production of color. Color was really due to commingling of both." In his Theory of Colors, von Goethe made an experimental analysis of the interpretation of interacting physical elements, such as the deformation/appearance of colors in a black and white world where shadow and light compete; this is a good example of how this appearance can be related only to the experience of the observer (Tyndall 2013; von Goethe 2013; Schopenhauer 2010). This situation is not much dissimilar from what happens in *Flatland* by Edwin A. Abbott, where the observations of the protagonist, who possesses the certainty of the existence of a three-dimensional world, lead him to manage this knowledge to interpret the reading of two-dimensional objects in a two-dimensional world (Abbott 2014).

Again the ambiguity of the perception in different dimensions described by Abbott reminds Plato's discussions about the impressions derived from shadows seen on the wall of a cave. Abbott gives the reader the task of understanding phenomena in a world with a higher dimensionality starting from the reading (vision) of slices with lower dimensionality. Perceptions, which are observations in the narrow sense, appear as theoretical constructions. Visual perceptions are often an unconscious construction of a mental three-dimensional model starting from two-dimensional retina *stimuli* that together with illusions appear in a persistent, metastable, cyclic manner independent from the acquired information (von Helmholtz 1896).

This is precisely the situation that scientists face as they analyze the samples' slices in electron microscopy images, in the attempt to reconstruct the objects' forms in the space by studying their planar sections—a process that implies lots of complicated factors, as illustrated by Claude F. Bragdon in *A Primer of Higher Space. The Fourth Dimension*. This is an issue easily understandable in the case of TEM that is also present, though in a less evident way, in SEM imaging (Mansfield et al. 2017; Bragdon 1913). The sample is dependent on both the energy of the primary electrons and their focusing, as well as on the type of the emitted electrons that generate the beam, secondary or backscattered. During the reading of images' data, the observer generates perceptions and illusions derived from the personal background learned from their earliest childhood and carried on throughout life (Schopenhauer 2010). The role of metadata is discussed also in Kandinsky's work devoted to the construction of a theory of painting, where the interpretation of the meaning of the representation of an object is stimulated and suggested by the unavoidable and necessary presence of the boundary (natural or artificial) (Kandinsky 1947).

What is experienced through the senses is felt as reality and should not be confused with the fruit of the mind, although reality and thought derive from the practice and elaboration of the same experience. For Aristotle imagination is a faculty that actively interprets the perceived data and stores, manufactures, and retrieves images, giving rise to a nonlinear process in which the observer is influenced by newly acquired information (feedback), with the result that who observes after is different from who observed before, one of the main features of quantum mechanics. "An image attracts, deceives, imitates, resembles, replaces, and animates. It is precisely this unruly behavior that renders an image so difficult to grasp" (Nielsen 2003). This observation should be kept in mind when facing the relationship between images (data and results of the experiment) and their interpretation. On the other hand, images are the means to grasp the results of an experiment, translating raw data into clear comprehensible language; images are also useful in establishing links with other similar figures derived from the everyday life and belonging to the mental archive of the observer/researcher.

Several years ago, when the understanding of subnuclear events began through the visualization of objects smaller than the eye could see and faster than the eye could follow, the LEP (Large Electron-Positron at CERN—Geneva, Switzerland) collider provided information about structures with dimensions in the order of 10^{-18} m and was therefore considered as an "ultra" microscope. In LEP collider a burst of energy generates hundreds of elementary particles monitored by hundreds of thousands of sensors. In less than a second, an electronic system must sort through the data from tens of thousands electron-positron encounters, searching for just one or two head-on collisions potentially leading to discoveries about nature's fundamental forces and elementary particles. When the electronic systems identified such a promising event, a picture of the data (a set of points in the space) was then transmitted to the most ingenious image processor, the most important resource in experimental physics: the human brain; in fact computers could not match the brain's capacity to recognize complicated and possibly improbable or never seen patterns in the data collected by the LEP detectors (Breuker et al. 1991).

This explains why it would be better to approach electron microscopy images with the same method applied to the analysis of a painting. Images, obviously not touched up, should give primary examples of the natural relation with the subject portrayed as it happens with paintings. The interpretation of electron microscopy images depends on the brain-mind architecture, and therefore, it is the object of study in neurobiology (bottom-up) and psychology (top-down), the latter aiming to understand the dynamics of thought through introspection.

The identification of an image content is typical of art and paintings as outlined by René F. G. Magritte, painter of unknown images of known objects, who thought that "Everything we see hides another thing, we always want to see what is hidden by what we see." Simply comes the correlation with our experiment, in which we have non-completely known images (due to their production process) of not so familiar objects, as polyurethane nanoparticles produced by the biodegradative activity of *Staphylococcus aureus*. No one perfectly knows the structures and functions of nanoparticles generated by biodestruction; as a consequence there is no prototype or blueprint as it happens for instance in the case of engineered nanoparticles for applications in the industrial or biomedical fields.

I learned very early the difference between knowing the name of something and knowing something.
Richard P. Feynman

The image encloses a synthesis of two ideas: the one implied in the image itself (i.e., what the operator wanted to search and explain) and the one of the observer with the personal burden of knowledge and experience. Hence images are not unambiguous denotative symbolic complexes, but rather they are connotative installations open to diverse interpretations (Flusser 2000).

Often when considering an image (static or extrapolated from a dynamic sequence) people tend to immediately report the informational content they see, assuming that a glimpse suffices to describe what is portrayed; actually an image holds a story in itself and only a thorough observation and the analysis of the spatiotemporal context can help displaying it. Again, it is helpful borrowing from arts the ability to inspect elements and fine details to learn how to investigate an image: images and paintings can be thought of as signs or tracks left by an imprinter (electron microscopes or painters) to be read by an interpreter (researcher or art critic) (Eco 2016). The semiotic relation between the parts, the initial idea of the author, and the content perceived by who looks contribute to the building of the technical features of the image (why and how it has been acquired) and to the recovery of the story (where and when) of the image (what), being the story a fundamental often neglected part of the technical features. The interrogatives, which find their source in Aristotle's Nicomachean Ethics, are fundamental for the collection of information and problem solving [Aristotle, ~350 B.C.]; they are constantly used in journalism and scientific and crime investigations concurring in the creation of a narration. Having its roots in the literary world, the narration has a major role as instrument for a deeper and smarter knowledge, shielding us from the impulse to reach a truth of the moment. Investigation should be the art of slow observation and should aim to the knowledge. Slowness protects us from hasty deductions derived by the view of a flawless reality, from the urgency of self-assertive investigation abilities, and from a selective blindness that would limit the observer's perceptive functions (Carofiglio 2019). The prototype of the investigator is Calvino's character Mr. Palomar: an observer who, as a modern Lucretius, devoid of any system describes and investigates nature starting from the elementary data of his daily experience, and that through narration tries to escape superficial and ready-made outcomes. Calvino's work highlights how narration can present an advantage above the deductive reasoning (Nicoli 2020; Calvino 1985).

Interpreting images means to extrapolate the significance of the image putting it into words; it is a translation from a visual imagery (image language) to current (spoken) language (Fig. 1). The observer's perception is ultimately biased since image interpretation is affected by personal experiences, cultural background and

Fig. 1 Visual representation by means of tags of some of the terms presented and discussed in this book. The image visually suggests the links and overlaps established by the use of synonyms previously discussed. "Extant nonnullae rationes, atque modi: quibus possunt tum nomina, tum res ipsae unica imagine figurari." "There are many methods and manners by which names and things can be formed into individual mnemonic figures." *Giordano Bruno*

education, as well as psychological processes. Although this condition could be seen as a disadvantage or even an obstacle, its richness of shades of uncertainty finds a lot of analogies with the reading of a text in its original language or a translated version.

As seen in the four different English versions of Pushkin's poem (Fig. 2) dedicated to science, human beings, and transcendental elements, translations of texts can be more than one, slightly or evidently different from one another, some more literally accurate, others less precise. Saint Jerome already in the fourth century faced this problem during his translation of the Bible from Hebrew and Greek to Latin. Realizing that sometimes the translation could not be perfectly loyal, and inspired by Cicero and Horace, he applied the *ad sensum* criterium, favoring the

So many wonders to discover
Are yet with the enlightenment spirit,
Experience, the son of painful errors,
And genius, the paradoxes' friend,
And accident, inventive god.

Oh, how many of wonderful discoveries
Enlightenment spirit prepares for us
And Experience, the son of difficult errors,
And the Genius, the friend of paradoxes,
And the Case, the God is inventor.

О сколько нам открытий чудных
Готовят просвещенья дух
И опыт, сын ошибок трудных,
И гений, парадоксов друг,
И случай, бог изобретатель.

А.С. Пушкин (1829)

How many wonderful discoveries
the spirit of education is preparing
for us!
Experience is like the son of
difficult mistakes,
and genius is a friend of paradoxes.

Oh so many wonderful discoveries
the Wisdom Spirit brings to us,
together with the Experience -
the child of hard mistakes,
and the Genius - the friend of paradoxes,
and the Chance - the Lord the inventor.

Fig. 2 Four different translations of Pushkin's verses found in different sources. The translation in the bottom left lacks the last verse; a curiosity about these verses is that they were the motto of the television program hosted by Sergey P. Kapitsa "Evident, but incredible" aired on the USSR television starting from 1973, which aimed at the popularization of science

meaning of the sentence over the literal translation word by word. The painter Michelangelo Merisi, known as Caravaggio, beautifully represented the difficulty of the translation task around 1605 in his paintings "San Girolamo scrivente" (Saint Jerome writing) in which the saint appears as an elderly man aggravated by the complex exegesis of the sacred text (Beian 2013).

The passage from an idiom to another can involve not only the transposition of a term in a different language, but sometimes it involves the alteration or transformation of the meaning as well. Thus the initial concept, as thought by the author, becomes a new concept reinterpreted by the translator. This is summed up in the Italian saying "*traduttore, traditore*" (translator, traitor).

A text can be misinterpreted or reinterpreted until the translation becomes treacherous, and the original *intentio auctoris*, the author's intention, gets either lost or muddled up with the reader's elaboration (*intentio lectoris*) which leads to an evolution (reshaping, remodulation) of the text and therefore to a partially or totally different conceptual meaning (*intentio operis*) (Eco 2016; Eco 1984).

A significant example of misinterpretation or reinterpretation can be seen in titles of famous works as in "Punkt und Linie zu Fläche," an essay by Kandinsky. The English translation *Point and Line to Plane* is loyal to the original one keeping the correlation of point and line to a particular type of surface; the Italian "Punto, Linea, Superficie" and French "Point, Ligne, Plane" translations distort this correlation promoting the background as a main character at the same level as the other two.

Another interesting case of mistranslation which involves differences starting directly from the title is the Russian into Italian translation of Fëdor M. Dostoevskij's "Двойник." The Italian translation "Il sosia" derives from *Sosia*, the Latin name of the character of the comedy *Amphitruo* by Titus M. Plautus. This translation loses the semantic value of the Russian word that contains the root "Два," which means "two." Quite different are the English and French translations, respectively "The double" and "Le double" which are loyal to the original title (de Vidovich 1984).

The reflection about translations helps in the comprehension of the difference between equality and similarity which is a key concept in image interpretation. Irreversibility and asymmetry are peculiarities of the translation process; indeed, though different translations appear to be telling the same content as the original, often they lead to slightly different meanings, making it impossible to recover the original text starting from the translated one. Hence the relation between translated and original text is not single-valued (identity) but a similarity. Generally, this multivalued relation is that of a term and its synonyms, which to a unique text correspond to a lot of translations, each one different from the others and with proper meanings.

The same issues described above play a key role in the intersemiotic translation between different systems, in the case of the world of images and their interpretation, from a visual to a verbal language. It follows that an unstructured and superficial approach is not prone to master the field of imaging. Especially in the scientific field it is necessary to avoid an approach based on "wild imaging" and interpreting facilitated by a lot of factors, from the availability of more and more massive storage systems, to the supply of machine time and experts for the collection and preparation of samples.

It's good to follow our intuition, as long as we make sure we aren't avoiding the work that's required to know whether or not our judgement is correct.
Garri K. Kasparov

Therefore it is mandatory on one hand to include also technicians and microscopists with their personal cultural background in the general idea of the experiment, and on the other hand it is important to assess the knowledge of researchers about the general and sometims detailed functioning of microscopes, so that following the steps of preparation and imaging of the samples, every one should be prepared about the kind of objects to presumably expect in the image.

In other words, you cannot observe a wave without bearing in mind the complex features that concur in shaping it and the other, equally complex ones that the wave itself originates.
Italo Calvino

Просвещение—Education

La ignoranza è la più bella scienza del mondo,
perché s'acquista senza fatica
e non rende l'animo affetto di malincolia.
Ignorance is the most delightful science in the world because it is acquired without labor or pains and keeps the mind from melancholy.
Giordano Bruno

To implement the skill of discerning objects in images it is of primary importance to have the knowledge, acquired through first-person experience, of the world of solids. Lack of this wisdom facilitates the misreading of images and subsequently leads to wrong or misinterpreted results.

Vision is not a proximal sense as taste, smell, or touch; thus it needs help from something else, from knowledge or past experiences to education or culture, for the interpretation of things. The main point is that each of these elements, being strictly subjective, is part of the personal luggage of every person (Gregory 2001).

Vision is a process that relates the subject (observer) to the object through a physical mediator: light. Its peculiarity is that, when it is separated from its source and detectors, light cannot be known, observed, or measured unless it interacts with some material object while traveling from its source to the object and from the object to the observer. In the frame of common education light presents the fascinating duplicity of behaving like a wave or a particle accordingly to the observer's experimental setup.

The complexity of light in general terms is so high that a philosopher and a scientist like Bruno discussing the phenomena of intelligibility in the sixteenth century created a theology of light based on optics and vision. Across centuries lots of references to light and vision were used when discussing about knowledge, from "illumination" and Enlightenment both in the Eastern and Western philosophical approach, to semiotic investigations about mirrors and the reflection theory, to the consequences on causality in the theory of relativity deriving from the most important feature of light, i.e., the constancy of its velocity that is unrelated to the velocity of both source and observer.

Knowing others is wisdom, knowing yourself is Enlightenment.
Lao Tzu

The topic of image interpretation is so important that in recent times two major scientific institutions, **CERN** (the European Organization for Nuclear Research in Geneva, Swiss Confederation) and **ALMA** (Atacama Large Millimeter/submillimeter Array in Atacama Desert, Chile), inquired about what artists and scientists have in common, and whether both ways of interpreting the world could merge into a unique language. Science, art, and image interpretation are strictly related; their synergy has made it possible to generate, after complex studies and operations, a picture (not a photograph but an image) of an unseeable object as the black hole in the Messier 87 galaxy, as well as images of the interactions of elementary particles obtained by LEP colliders (Cooper 2019; Henry, 2014; Bello n.d.).

The concept and processesof image interpretation are well illustrated by an image from the book *The Little Prince* by Antoine de Saint-Exupéry, where what at first sight it appears to be a hat when it is actually a boa snake which ate an elephant (de Saint-Exupéry 1943). Of course the comparison between the biased mentality of an adult and the limitless Aristotelian imagination and naïvety of a child is very suggestive. This example reveals how the perception of a person is pliable and guided by everyday experience, so that the thoughtlessness of a child and his fancy explanations can disarm the mental walls that adults build during the years. Seeing a pictured object and subsequently defining it could seem a simple operation, but actually it is an enterprise which requires a whole hierarchy of functions. During the vision objects are not seen in their wholeness, but fragmented into "forms, surfaces, contours." People do not even realize that this process is happening, but it continues working in different contexts and conditions, extracting as many information as possible and storing them in our background. The interpretative mechanisms of images could be better understood studying how blind people manages to "read" Braille images, where the visual automatisms and specific experience of the vision are replaced by the feel. The process has also a lot of shades and differences if the subjects interpreting images are people born blind or people who experienced sight problems in advanced age; in fact the approach to the world of objects learned through just the manipulation of things is totally different from the sensation felt through touching paired with a visual (even partial) experience (Sacks 2010). Moreover what a person experiences seeing or in this case touching a picture is extremely unique since the sensations evoked are closely related to a person's own past events, culture, and background. Curious is the case of Borges whose blindness did not prevent the writer from starting to draw, even a self-portrait, visit Sicily and appreciate its historical sites (Luque 2017).

Pictures show more than cannot be said in words; thus their transposition into a verbal language is not an easy, objective, and unbiased task. What is applicable to pictures, photographs, and art in general is *a fortiori* valid for electron microscopy images and is a key topic of this journey which aims to look for the unseen and unsaid about images in a scientific context.

Art forms like texts and pictures can be read in different ways (Barthes 1980): writers using punctuation suggest pauses that help the reader to interpret the text, whereas painters and photographers help the viewer to interpret a picture through the use of shadows (absence of light) and perspective. Shadows in ordinary photography correspond to a lack of signal, thus to a white point or pixel in the negative and to a black one in the final photograph. The process is different when talking about digital photos and micrographs; in the case of electron microscopy the link of the object to the resulting picture requires a more elaborate sequence of steps: electron bombardment, electron collection, electron transformation into photons, signal multiplication, and electronic/digital photon acquisition.

The validity and interpretation of computerized, machine-made images, derived as in our case from electron microscopy, should always be questioned since it would be a mistake to put the tool makers in the privileged position of deciding how their tools should be used. This was one of the starting point for this book on electron microscopy: the role of images in the formation of knowledge and the feedback role of knowledge in the interpretation of images. The same issue was faced when scientists working on the tracking images of elementary particles affirmed that physicist could not rely completely on computers for data analyses, since computer automatically analyzing events would be limited by the programmers' expectative. This kind of systems risks to selectively delete information, forcing the interpretation to remain in the world of the ordinary already known phenomena. This has to work as a reminder that the past and the present are connected, and as a warning against the potential tyranny of the newest digitized images that, though often beautiful and beguiling, are anyhow man-made and prone to mistakes. The same problem can be posed when the electron microscopy images automatic analysis and data retrieval is at stake. "The more technological the image looks, the more it exudes ... authority, but a computer is, nonetheless, a man-made tool that seems to promise a non-human precision" (Crawford 2021; Holtzmann Kevles 2007).

Often it is assumed that photographs (film and paper, negative and positive, prints or digital and monitor reading) obtained by light exposure or electron microscopy are absolutely comparable. Actually, when exploring the use of scientific tools it is important to know what they exactly do, so that one should know the instrument used, its behavior, and the nature of the imprinter (light or electron beam).

Simple elements of modern physics highlight that the abovementioned sources are very different: indeed, photons are mass- and charge-free particles with a constant speed in vacuum (the maximum measurable speed possible for any entity on the basis of the general laws of electromagnetism and the theory of relativity) obeying, as identical particles with integer spin, to the statistics of Bose-Einstein (Tarasov 1980; Davydov 1976). Electrons have a different behavior since they are massive and charged and can be differently accelerated to reach various velocities as can be seen in the recorded parameters of an electron microscopy image, although electrons will never reach the speed of light. Electrons cannot compenetrate since they obey the Fermi-Dirac's statistics and therefore the Pauli's exclusion principle applies. Despite these differences, both electrons and photons are the inextricable protagonists of electromagnetism and therefore the actors of one of the three fundamental interactions used to describe natural phenomena, the electroweak one (the other two being the strong nuclear and the gravitational ones). Thus, although different from each other, they communicate and interact so deeply that, as illustrated in the case of the free electron laser, the change in the state of electron motion, obtained by electromagnetic means and already known for more than one century, is associated to X-ray generation where electrons can generate photon beams or rays of light. This phenomenon has more direct manifestations, such as the Bremsstrahlung or "braking radiation" effect (electromagnetic radiation produced by the deceleration of a charged particle): the moving particle loses kinetic energy which, because of energy conservation, is converted into electromagnetic radiation (photons generation).

Going back to the topic of interpretation, other examples can be borrowed from the arts. Commonly a language is based on sounds and on their interpretation, mediated by education and environment. "Grammelot" is an agramatic stage language that is not based on word articulation, but reproduces some properties of the phonetic system of a particular idiom, so that what is actually a continuous stream of quick and arbitrary sequences of sounds resembles to all effects a speech. It has a strong expressive mimic-gestural component that the actor performs in parallel with sounds, giving the audience the real impression to listen and recognize words that do not have a proper meaning. The first form of grammelot is used by children with their incredible fantasy when they pretend to make clear speeches with extraordinary maulings. Though in Italy grammelot has illustrious antecedents in the multiple dialecticism of the Renaissance comedy and in the Commedia dell'Arte, its creation is the work of Dario Fo who invented languages articulated in a semiotic code (Bandettini 2014; Fo 1997). A parallel example in the field of painting can be found in the activity of Giuseppe Arcimboldi better known as Arcimboldo, an Italian painter notorious for his burlesque portraits, or grotesque physiognomies obtained through bizarre combinations of fruits and vegetables, fishes, birds, books, and vases. Arcimboldo's works are clearly based on the artistic trickery of double images; Galilei expressed critics about these paintings saying that the "oblique" mentality of the artist was expressing itself in double images of this kind (Berra 2017; Panofsky 1956). Actually duplicity is a theme present in art and science since the very beginning.

Semiotics is the discipline of languages that through keywords such as print, track, imprinter, and impression describe the manifestations of the natural world and the cooperation between the brain and the eyes in image interpretation. The starting point is that images are tracks, light waves, or photons imprinted on an emulsion (film) or on a grid of semiconductors (CCD camera).

A photograph is not just a representation, but an acquisition of experience through various grades of knowledge. It is a record that can be accessed at different times. In this way it has two main features: the possibility of reading and re-reading (with implication that will be discussed later) and the duration (natural or artificial) that can be limited or infinite depending on the expertise in material sciences (where the material is a physical base of recording) and in storage systems architecture. Both these features show that for digital storage of pictures and more in general of data, time has a fundamental role and is at the basis of memory, a building block of experience and knowledge. Knowing means acquiring information that are elaborated and transformed. Photos as well as paintings are means of knowledge communication associated to the transmission of information, under the form of narration. In science, communication is a basic element and the verbalization or description of results (narration) can be done by the observer or constructed starting from the descriptions of other observers.

Linking science with art, literature, and photography, the observer (the spectator of a painting or a photo, the reader of a text) interacts dialoguing with the operator (the microscopist, the painter, the photographer, or the author) in a dialectical engagement through which the reasons of a scientific investigation, a painting, a photograph, or a text are defined. The observer and operator that sometimes are the same person can have different or common purposes. Often the observer reading a painting, a photograph, or a text, positively or negatively biased by his own experience and knowledge, projects something that does not exist, so that eventually the image or the text does not correspond to the initial intentions of the author (*intentio auctoris*) (Eco 1984). Moreover this projection of nonexistent things makes them real, so that the narration gains an additional role of active producer of data, and as a consequence of facts, and various points of view and intellectual perspectives are introduced.

Thus as it happens for translations, the interpretation of paintings, photographs, or texts has not a biunivocal correspondence between the author's idea and the interpreted versions. Filippo Brunelleschi, one of the major personalities of the Italian Renaissance architecture, well aware of this possibility had the genial idea of a pinhole (*foro stenopeico*) to force the observer to select one specific point of view among the many possible, giving a solid basis to geometrical optical linear perspective (Kemp 1999; Kemp 1992).

What has been told so far has some implications in the activity of the observer for the identification process, i.e., the building and knowledge of the identity of the *spectrum* (the target, referent, the eidolon of what is going to be narrated) chosen by the operator. These sociological implications apply to human beings as well as to matter components. The identity of an individual is not self-consistent but has to be built in relation to the environment (the others). This originates two identities: "what one is and what one really and truly knows" or "the difference between what one is and what one thinks one knows but does not really know." This sentence is a valuable affirmation especially when working in the nanoparticle field, since the focus of the problem is the identification of nanoparticles. Calvino's brilliant idea of explaining the advantages of reading classics is a precious hint for researchers to not only rely on novelties of scientific papers but also to read and re-read the classical scientific textbooks in order to consolidate knowledge (Amodeo 2013; Calvino 2009; Lynch et al. 2009).

Another example of the intriguing complexity of the role of narration as active producer of data can be seen in the 1917 play *Right You Are (if you think so)* by Pirandello. In this work the spectrum of a character (Signora Ponza), refusing her identity, assumes antagonistic personalities that others confer on her, thus becoming another spectrum with a proper life but driven by the environment and devoid of self-knowledge. This is brilliantly resumed in the conclusive line of Signora Ponza: "No, for myself…I am she whom you believe me to be" (Pirandello 2020). What represented by Pirandello in his comedy can be thought of as a forced representation of real life, but 80 years later in the same geographical location it happens that news anticipate facts and manage to create doubles of relevant people leaving traces in procedural documents (Parlagreco 1990). Similar plots appear also in the cinema where a multiplicity of possible individual "points of view" leads to misunderstandings or dramas, as for instance in Pier Paolo Pasolini's "What are the clouds?", third episode of *Caprice Italian Style* (1967) inspired by Shakespeare's Othello.

> *This was the great truth of life, that fact and fiction were always merging, interchanging.*
> *Graham Swift*

The fuzziness is also incremented by the relation of the text with the reader due to the translation that in some cases can lead to the distortion of the *intentio auctoris*. As already mentioned in the conclusive line of Pirandello's play, Signora Ponza says: "No, for myself…I am she whom you believe me to be" or as translated by Arthur Livingston in 1922 "No, for myself I am…whoever you choose to have me." The distortion between "to be" and "to have" strongly recalls the transition of the protagonist of Calvino's "The adventure of a photographer" from a sentimental relationship to the possession of the partner (Pirandello 2020, 2015; Chiappori 2014; Calvino 1984b).

> *Taking pictures means establishing with the world a particular relationship that gives a feeling of knowledge and therefore of power*
> *Susan Sontag*
> (Authors' translation)

Calvino's story rotates around the theme of identity with the aim to answer to the primordial and elementary question "who am I?" that is connected to a series of unexpressed and returning questions such as "who am I for others?, who am I for myself?" (Scarpa 2005).

The questions can be transposed in other fields with a symmetric relation: "who are you?" or "what is this?" or "which object can be associated to a defined set of black and white points/pixels?" or even more critically "which object do I select among the wealth of possible objects associable to them?". These more and more complex articulations actually do have consequences and each one of them causes a progressive slowing down in the decisional skills of the observer.

This is illustrated in the FIB/SEM process of image acquisition when the operator is in search of the best point of view for the best image of the object. The problem is amplified by the availability of two beams (ions and electrons) and several detectors based on different detection systems, and by the possibility that at times the operator is substituted by an automatic detection system (with its pros and cons).

A skill that sharpens the explorative will and ability of the observer is represented by the intuition. Modern (quantum) physics, studying the laws ruling natural phenomena, the interactions among matter constituents, and the relationship between the world and its representations, deeply discusses this topic. The importance of the observer's role is underlined also in literature for instance by Carlo E. Gadda, an Italian writer, who in his personal view considered knowledge as a means to insert something in reality and hence distort it. This is easily seen in electron microscopy investigations where in order to be explored, the sample is distorted by the tools (electron and ion beam) necessary for the image building. The considerations made for text-based descriptions and knowledge must be taken into account by those who try to know objects and understand processes via images, leaving in the background the controversial problem of giving a definition of "object".

The connection between science and different forms of art appears always more intricate. The theme of the double present in literature, but common also in visual arts, introduces bifurcation as a feature of character observation. As an example, Borges in "The garden of forking paths" (1941) illustrated the concept of bifurcation and "route to chaos" years before the term bifurcation massively entered the fields of physics, synergetics, and statistics of irreversible processes (second half of the twentieth century) as an indicator of phase transitions and one of the key mechanisms that by iteration can lead to chaotic behaviors (Borges 2018). The chaotic behavior appears as more than a simple assonance between narration and physics of irreversible processes, since the same Calvino in a short story of 1955 described the progressive appearance of madness (chaotic behavior) in the protagonist (Calvino 1955). In a chaotic behavior small uncertainties (fluctuations) at an initial state can lead to large errors in a final one; even simple systems for which all forces are known can behave unpredictively. A well-known example and the basis of any discussion on chaos is the problem of long-term weather forecast (Motter and Campbell 2014; Kadanoff 1993).

A duplicity view materialized in geometrical perspective requires that the observer combines two different fixed views of the object; thus the observer transfers the double view of the object to himself, realizing the own doubling and splitting of the self. This is well represented by the duplication of the self that appears when the observer has access to the self-image generated by a mirror. The splitting of self is rendered by Dante in "Vita Nuova" (1294) where the author triples himself into a narrator, a poet, and a scholar, actually building a text where he can both write and read the self in multiple ways (Hoskin 1995). In paintings duplicity appears in *Self-Portrait with Mirror and Easel* (1646) by Johannes Gumpp, an Austrian artist of the seventeenth century, where a self-portrait is associated to a portrait of the artist as he sees himself in the mirror. In general, a self-portrait is a representation of an artist that is drawn, painted, photographed, or sculpted by that artist; hence the self-portrait, being an approximately faithful reflection of the world in consciousness, produces knowledge.

More in detail knowledge, the aim of this research activity on nanoparticles based on electron microscopy, can be seen as information stored in the mind and in human artifacts that forms the basis for both the human thoughts and actions. Knowledge has an intrinsic duality: the knowledge of the object of the investigation and the knowledge produced during the process of knowledge seeking. Hence knowledge is built from and at the same time builds education, implying the construction of experience that at times can be pointed by mistakes (Bhaskar 1978).

These elementary and anyway basic considerations allow us to connect the artistic expression of Pushkin's poem to our actual problem of understanding the formation process of nanoparticles, linking experience, education, and mistakes to theory and practice in science. The most direct tool to acquire knowledge about nanoparticles is via electron microscopy imaging; this process is the result of a nonlinear set of theoretical and practical couples elements like observation, description, education, etc.

Experience is based on the observation of objects and their behaviors; observations give proofs of the existence of objects and their related phenomena. There is a direct but complex link between proof and phenomenon, and one has to be aware that if there is no proof, consequently there is no phenomenon.

> *Absence of proof is not proof of absence.*
> *William Cowper*

The first element of this complexity derives from the strategic role of the observer which actually interacts with the voluntary or involuntary observed system; the second issue is that "no elementary phenomenon is a real phenomenon until it is an observed phenomenon" (Dolling et al. 2003). This sentence hides problems and contradictions evident even when associated to non-scientific fields; can the lack of proofs that a fact actually happened be considered itself a proof that the fact never happened? According to the reasoning just described, the proof tells the truth about the phenomenon. This provokes another reflection about the questions one asks. If the questions are imprecise and confused, they can only lead to imprecise and confused answers; if these misleading answers are taken for good, they are to be considered as lies (Black 1964).

> *We can't define anything precisely. If we attempt to, we get into that paralysis of thought that comes to philosophers... one saying to the other: "You don't know what you are talking about!"*
>
> *The second one says: "What do you mean by know? What do you mean by talking? What do you mean by you? What do you mean by know?"*
>
> *Richard P. Feynman*

When we explore matter by ever-expanding magnification, into a past infinity, and collect images of faint objects so small, what exactly are we seeing? Maybe the things that we see are different from reality and far from the truth.

> *We do a lot of looking: we look through lenses, telescopes, television tubes...*
>
> *Our looking is perfected every day, but we see less and less.*
>
> *Frederick Frank*

Often electron microscopy images are blurry, distorted, and riddled with background noise. In order to see with any clarity, these images are finely combed over with artificial changes; due to this process it becomes fundamental to have an *a priori* knowledge of the object under investigation. "Creating an honest image from electron microscopy data is as much an art as a science. When processed correctly, an attractive and evocative picture brings out the scientific content within." There is so much manipulation that accompanies this process that it can be questioned whether what we are seeing is in fact the real deal, a surprisingly complex question that leads to sometimes embarassing uncertainty (Peters 2017; Villard and Levay 2002).

In black and white electron microscopy images, colors are not present as a source of information and should be excluded. Image analysis is usually based on a variety of parameters, among them size, shape, site, association, shadow, and pattern. In some cases, a single element alone is sufficient for a successful identification; in others, the use of several different elements is required. Black and white photographs show that light alone conveys information. In analogic images the white dots have an evident origin, deriving from the knowledge of the photographic process used to capture the real object, whereas in digital images, in particular in those generated by an electron microscope, the roles of pixels change over time (earlier associated to the presence of material, then to its absence). Bearing in mind that in some cases the black and white perception associates different meanings to the resulting shapes, even this mechanism of image building can result in misleading and ambiguous outcomes. As Plato knew and as it is evident in shadow puppetry, even the absence of light, i.e., the shadow, is an instrument of knowledge: revealing two-dimensional shapes, shadows suggest the existence of a variety of three-dimensional objects that cast them. In other words shadows could be interpreted in various ways, all linked to concrete objects, even though the shadow is actually abstract, unstable, and uncatchable. This can be especially seen in TEM and STEM images because they are the recording of the real shadow of the object shielding the electrons from the source. Trying to answer the key question "what are nanoparticles?" we should avoid apodictic statements about something we do not fully know, given the impossibility of answering with certainty to such a question. Lots of intellectual revolutions were achieved posing questions about apparently obvious definitions that instead were the fruit of traditions, habits, or mistakes (Malvaldi and Leporini 2014).

The correspondence between a bidimensional entity and a multiplicity of three-dimensional objects is at the base of the ambiguity leading to deceiving processes. The etymology of the term "ambiguous" highlights that something can be intended in two or more senses, generating either doubts or deceit. Therefore, shadows being imperfect copies of real objects recall once again duplicity, a theme related to the problem of doubleness and identity, and well developed in literature, art, and science (double, mask, mirror) through the centuries. As an example, the theme of Rachel and Leah, characters already known through the Bible, is presented in Dante's Purgatorio (Purgatory, 1320) in the form of dichotomy as the two routes to knowledge, each with a mirror in hand symbolizing respectively the theological and the empirical truth (Alighieri 2014).

The mirror is a common tool strictly associated to images and perception. Already more than 2000 years ago Saint Paul mentioned the mirror in the first letter to the Corinthians; later Saint Thomas Aquinas linked speculation to the "*speculum*" stating that seeing something through a mirror is like observing a cause through its effect. The theme of the mirror is also present in Dante's *Comedia* (Divine Comedy), in the second chant of Paradiso (Heaven) where Beatrice proposes an optical experiment to the poet: "Taking three mirrors, place a pair of them / at equal distance from you; set the third / midway between those two, but farther back. / Then, turning toward them, at your back have placed / a light that kindles those three mirrors and / returns to you, reflected by them all." Then Beatrice arranging behind the pilgrim a source of light realizes an optical system intended to refute the erroneous ideas of Dante about the spots on the moon's surface, anticipating by three centuries Galilei's experimental activity (Melchior-Bonnet 2002).

Sometimes image analysis can give information about facts that are not the direct object of investigation and that can be useful to prove or disprove certain non-obvious propositions. For instance, observing the picture of a monument, known the height of the sun at the time the photo was taken, trigonometrical calculations give the height of the monument. If the height of the monument is known the process can work in reverse to estimate the date and time at which the image was acquired (an identification that is not necessarily univocal but that, again, establishes a link between photograph and time). A famous example that well illustrates this process regards the arctic mission conducted by Robert E. Peary in 1909 and supported by the National Geographic Society. Thanks to the lengths of the shadows visible in the photographs Peary took, in 1989 it was proved that the photographs were taken within 8 Km from the North Pole, granting that Peary accomplished the mission (Leary 1989).

Information can be retrieved also from images acquired in an unusual and involuntary way, as in the case of the nuclear bomb explosion of Hiroshima (1945) in which the nuclear flash imprinted a shadow on the stone steps of the main entrance of the Sumitomo Bank, likely suggesting the presence of a person waiting for the bank to open. Actually the nuclear flash acted as the flash of an ordinary camera, imprinting the "graph" (intended as its Greek meaning of "to write, draw, carve") directly on the stone, as documented by a photo of the US Army (Lévy-Leblond 2007). The last example, dealing with light and evidently matter, has analogies with what happens during the buildup of an electron microscopy image (micrograph). To fully comprehend this passage it is useful to think of the process with the terms of semiotics, where the impressor imprints the image. The resulting artificial or natural track (image) can have more than one simultaneous impressor, so that the role of particle and radiation in the nuclear flash (even though of different nature) is comparable to that of electrons, both primary and secondary, associated to the electron beam.

The image is read as a parallel or serial sequence of black and white pixels, the black ones associated to the absence of light or to shadows; in CCD cameras or electron microscopy images the relative role of blacks and whites can be inverted, depending on the number of physical processes required for the generation of the final trace. Limiting ourselves to the serial sequence of black and white pixels, interesting considerations for the cognitive value associated to the image are discussed by Calvino in "Mr. Palomar" about the songs of blackbirds. If we consider the single building blocks of the birds' song, i.e., each call, we can associate the call to a white pixel and each time unit of silence to a black one. On the basis of this analogy, we can compare the two languages, that of images (with the simplest gray scale) and that of the birds, counting each white pixel or single call as 1 and each shadow or unit of silence as 0. This analogy can be useful to evaluate the acquisition of scientific knowledge via photography and electron microscopy. First of all, we must answer two questions: (1) who do we take photos for? and (2) are information resident in the white pixels or in the black ones? Mr. Palomar in "The blackbird's whistle" asks himself "is it a dialogue, or does each blackbird whistle for itself and not for the other?" and "what if it is in the pause and not in the whistle that the meaning of the message is contained? If it were in the silence that the blackbirds speak to each other?" (Calvino 1985). "Mr. Palomar always hopes that silence contains something more than language can say" as if saying that black pixels (shadows) have in themselves more information or suggestions that one could expect—a strand well illustrated by famous writers as Christian Andersen in "The shadow" (1847), Guy de Maupassant in "Lui?" (1883), Adelbert von Chamisso in "The extraordinary story of Peter Schlemihl" (1814), and von Goethe in " Faust" (1808). Calvino's sensorial observations of the blackbird' song, based on the concept of exchange of information, are supported by scientific works that analyze the musical aspect of blackbird' songs with sonograms which can be read and re-read in search of structures and patterns (Piana 2007).

Usually the identification of objects and structures in images is performed by people according to different degrees of freedom. Digital processing and analysis could be carried out automatically to extract information but only in rare cases it can utterly replace a human interpretation, since subject-related influences (experience, education, and knowledge that contribute to build involuntary or associative memory) will inevitably vary the results, sometimes adding value to the meaning, creating illusions, or opening new frontiers to the interpretation enhanced by the ambiguities present in the images (Zeki 2004). Substantially, one of the ways to know what a nanoparticle is in a biomedical environment, is to face the issue of divisibility. Nanoparticles are aggregates of atoms/molecules, in forms often different from macromolecules, and have different properties which allow them to interact in different ways with living systems. It must be posed a preliminary question: when does an aggregate of atoms become a nanoparticle? What resembles the real answer can be only given by the first system that interacts with the unidentified aggregate, i.e. the cell with the role of observer. Anyway we have to keep in mind that the researcher with its cultural background, experience, and phantasy in fact is the ultimate decision maker of what the cell sees.

Images, photographs, and shadows are tools at the base of the pursuit of knowledge. According to the thought of Nicholas of Cusa "knowledge comes through similes": knowledge can be achieved creating an analogy (simile) between different objects (entities) which however have some common qualities or properties in a sort of "comparative likeness" (Cusano 2003). An example of knowledge acquired through similes is establishing a parallelism between a nanoparticle and a sea wave as investigated by Calvino's Mr. Palomar. It is not knowing all the single details that we really understand a phenomenon, but knowing in detail how it globally works; taking into consideration only the single atom or the unitary constituent of the wave without correlating them to the contest is useless to understand the process of a nanoparticle or a wave formation. This is well illustrated in different domains of synergetics, such as laser physics or oscillatory chemical reactions. Mr. Palomar's experiences always focus on a phenomenon isolated in space and time, following a reductionist linear approach in contrast with nonlinear collective processes, fundamental in synergetics and dissipative structures. The dynamics of Palomar's way of reasoning seem to follow a scientific method, but actually there are a lot of omissions and approximations, which make the approach unsatisfactory even for Palomar himself. Referring to the example of the sea wave, it is clear that a wave per its own definition cannot be divided into its unitary components; otherwise it would lose its main feature: periodicity. Most people, well represented by the character of Mr. Palomar, assume that a wave has a simple and univocal definition, but "defining this basic concept isn't so easy." Mr. Palomar does not manage to bear in mind all the aspects of the problem at once; hence his approach elides space and time organization and periodicity at the base of the wave's definition. This induces Palomar to lose his patience and abandon his research, but what if he took his observation as good and went on with his unreliable investigation (Scales and Snieder 1999; Calvino 1985)?

In theorem XXXVI Euclides relates the visible size of an object to the distance from the eye and investigates the shapes of cylinders and cones when viewed from different angles. Euclides proves the "the wheels of chariots appear circular (κυκλοειδής), sometimes pressed in (παρεσπασμενοι)." As Leonardo da Vinci wrote, on the basis of his own experience, masters do not trust the judgment of the eye because it can be deceitful (Knorr 1991). Following Leonardo's philosophy of the eye as the path to the true knowledge, in our investigation we dissect the electron microscopy images, keeping in mind that to see what is actually in front of us is a rare skill, biased by past experience, thoughts, or indoctrinations. It is important to not forget that a picture is a report on nature and that its interpretation will not be complete without answering to some of the questions listed in the Aristotelian Nicomachean Ethics (Raepple 2009).

Any representation is linked to its subject; this is due to an isomorphism or a homomorphism between an object and a subject. The correlation can originate from different sources such as a causal story, embedded information, the producer's intentions, but also illusions and ambiguities. Once again we will again use the help that the relationship between art and science can give us, focusing on the image, trying to formulate the conception of art as "imitation of reality" (mimesis), based on the concept of resemblance. Art and the creative process become for the scientist an explicit and direct way to experience that results no more undermined and confined (Zeki 2004; Gombrich et al. 1972; Black 1964).

Shadows, as seen in Plato's philosophy as well as in transmission electron microscopy, trigger the imagination creating illusions (departures from the truth) and reality distortion, such as ambiguities, paradoxes (wave-particle dualism), and fictions (time travel). But if reality and illusions are similar, the question of distinguishability and identity needs further discussion.

Similarity connects reality to sensation and is therefore a fundamental concept in building knowledge; in addition it is useful to sort objects in classes, even though the process is debatable. When thinking of a nanoparticle, due to its smallness, one of the first associations that comes to mind is the one with a geometric point. So the nanoparticle is ideally projected as a little, spherical object considered indivisible and therefore acting as an elementary character. This recalls the process of an ideal common representation of an electron, imagined as a point with an evident approximation due to the nonconsideration of the extension of its electromagnetic field or again the ideal representation of a photon as a point in the space even knowing that it is not possible to determine that precise point in a precise time.

The boundary between imagination and perception is not so sharp, as represented by Kandinsky that depicted ten different shapes of an object commonly intended and described as a tiny dot, pointing out the contrast between the thought and the perceived form. The problem of the point and its forms, absurd in Euclidean terms, was already raised by Bruno in the sixteenth century on the basis that abstractions nontranslatable in images would found on incommunicable concepts (Bruno 2015; Kandinsky 1947). The same thinking process can be applied when investigating nanoparticles, so that elements as the concept of exterior, size, and shape become fundamental (Fig. 3).

In the case of electron microscopy images the sectioning plane is the focal plane generated by the focusing of the electron beam. So instead of having an object, the simplest representation being a cube, that meets the plane and gets sectioned (Bragdon 1913), we have a plane that moves towards the target and slices it. Nanoparticles have

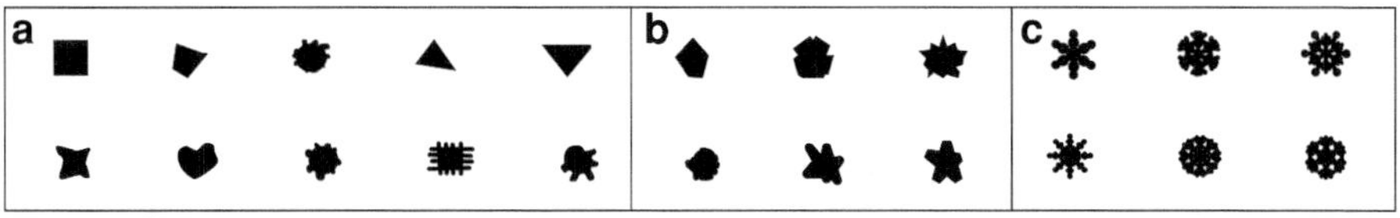

Fig. 3 Different forms of the point. (**a**) Reproduction of Kandinsky's "Examples of point forms" (Kandinsky 1947); (**b**) different representations of point forms as drawn by the authors; (**c**) illustration of different forms of ice crystals. In the real world the point presents itself under an infinite number of forms. Ideally the point has no dimension and often it is associated to a circular shape. Intriguingly the circular shape does not appear in Kandinsky's figure (**a**). Similar considerations can be applied to electron microscopy images of nanoparticles at different magnifications

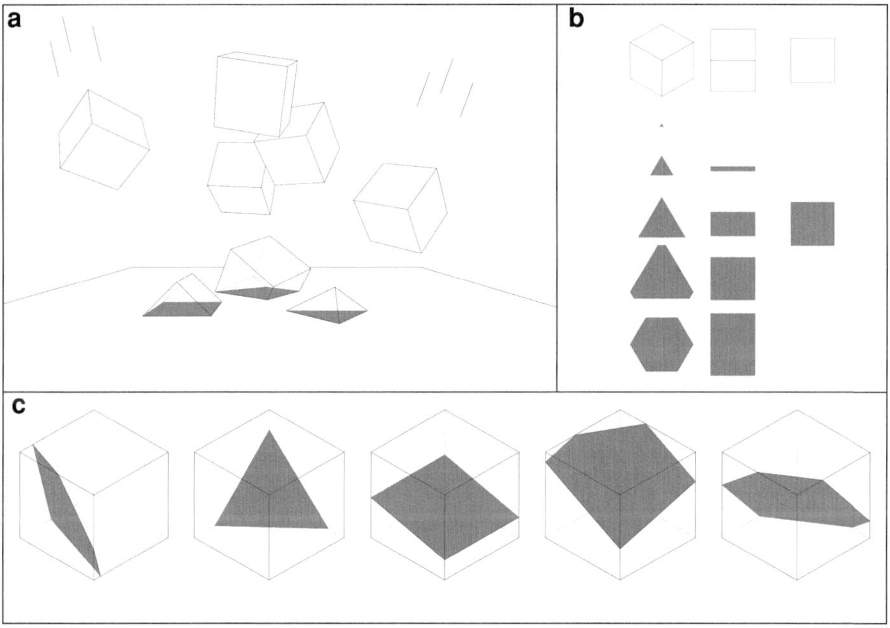

Fig. 4 Different forms of nanoparticles seen following the operation of slicing a cube (Mansfield et al. 2017; Bragdon 1913). (**a**) Overall view of the possible results of the interaction of nanoparticles (schematized in the simplest form of a cube) with a plane (in transmission electron microscopy the focal plane is in movement). (**b**) Classification of different possible outcomes of the sectioning operations. (**c**) Representation of how the slicing can affect the observed form of nanoparticles. *"L'instant engendre la forme, et la forme fait voir l'instant." "The instant generates the form, and the form shows the instant." Paul Valéry* (Authors' translation)

a variety of shapes that are modified by the interaction with the electrons, but as a consequence of the structure of the electron microscope, each nanoparticle can be seen as composed of a variety of elemental cubes, each giving rise to one of the possible sections (and of course this could be connected to the wealth of discretization processes that take place in the generation of the electron microscopy image) (Mansfield et al. 2017) (Figs. 4 and 5).

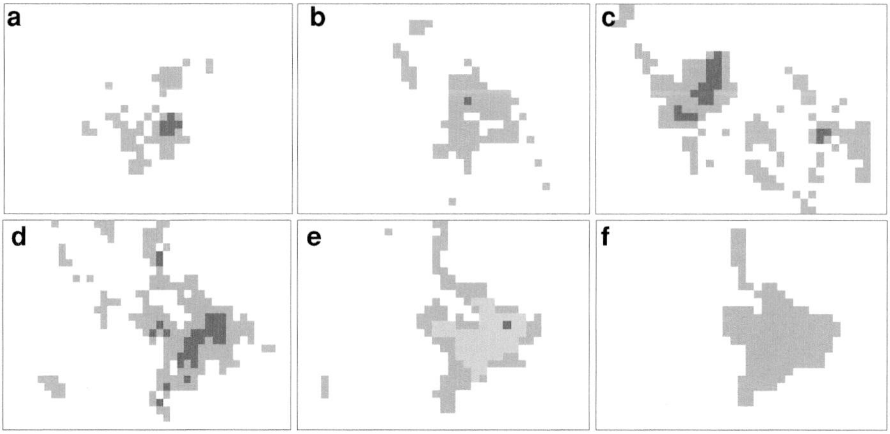

Fig. 5 Shapes of nanoparticles. Effects of the electron beam on nanoparticles during image acquisition and data manipulation of the acquired image influence the perception of the shapes of nanoparticles. (**a**, **b**, **c**, **d**) Analysis of four nanoparticles belonging to the same membrane vesicle. (**e**) First smoothing of image. (**d**, **f**) Second smoothing of image d (Erega et al. 2017). Smoothing a data set allows to catch important information, avoiding noise or fine-scale details; after a second operation of smoothing the result is stable and more similar to what a nanoparticle is expected to be, although the smoothing is a manipulation of data

Kandinsky's works on the nature and the properties of the fundamental elements of the form, the point and the line, establish connections between them and to the necessary support for their existence, the surface. The natural environment in a painting is a two-dimensional world: it is the canvas that at the same time is environment and support. The point when materializes must have a certain size, occupying a portion of the environment; it must have certain contours, which separate it from the surroundings but at the same time connect the point to the surface. The point can be defined as the minimum elementary form, but it is difficult to establish the exact limits of this minimal form, which depending on the observer's activities can grow and acquire surface. So the question about the dual aspect of a point as both a point and a surface with the passage to different size scale should be raised. The process of transformation of a point into a surface can be facilitated when a thin line appears next to the dot.

For Kandinsky the form, in all its species—natural and artificial—is a manifestation of a reality that can be contextualized only in relation to its background. For instance information about a point can be extracted only thanks to the presence of a line, so the line, when associated to a point, acquires an interpretative role. Also the line, traditionally defined as "length without thickness," must be concretely visualized in its own extension. In Kandinsky's view a line is a moving dot, so that every issue associated to the point can be applied to the line as well.

Kandinsky well illustrates the possible mental processes active in the study of nanoparticle images.

In the case of the observation of a membrane vesicle containing polyurethane nanoparticles, the line's role is played by the membrane of the vesicle that helps the observer in the identification of the object in the environment. In this way the nanoparticles, considered as points, can be seen inserted in a surface, the image. Moreover in the biological medium processes evolve over time so that shapes, sizes, and properties change constantly, as shown by the dynamics of the formation of the protein corona and the aggregation of nanoparticles, with substantial modifications regarding not only form and dimension but also the electromagnetic properties of the objects.

When looking at their interactions and considered not for what they appear in the image, but for what they mean for our knowledge, nanoparticles play a completely different role. Moreover, in order to catch hidden features, the focus can also move from single nanoparticles to the general view of the bacterial biodegradative activity. In this way image and concept become an inextricable ensemble (Cassirer 2009; Marramao 1992).

The thought, synthesis of the knowledge and at the base of its transmission, is the result of cascades of approximations interspersed with doubts. One of the forms of the transmission of the thought, without which science would not exist, is the narration. Narration is not only a vehicle of data but has also an active role in their production, reinforcing the value of approximations and introducing the simultaneous presence of truths and falseness. Added data and the connections created during their elaboration are not necessarily part of the object, but are voluntarily or involuntarily reshaped or distorted by memory and imagination. This is why the figurative representation of the concepts not only makes them communicable, but it is also of great help in determining peculiarities that in a verbal narration would remain neglected or indefinite (Bruno 2015).

Images may be uncomfortable or inconvenient, but only by embracing rational thought can we find proofs to build the basis of scientific analyses. In light of this, are images facts regarding a phenomenon or are they a proof of the phenomenon itself? Surely the question raises confusion and probably even illusions.

Another disturbing factor in the real world is that the reasoning process often starts with certain *a priori* and works back to find facts that support such preestablished conclusions often dismissing those in contradiction (Jones 2016). The issue finds its roots in Plato's distinction between eidos (idea) and eidolon (simulacrum) and the Aristotelian eikon (true image). People are biased, so it is extremely difficult to excerpt objective reality from images corrupted by subjective perceptions (Cassirer 2009). Images, even original shots free from any post-production operation, do not always represent the truth and can tell different stories, sometimes biased by a sort of "truthiness," a convenient truth not supported by facts.

Hence the arranged story narrated by an image may be used to mislead, create ambiguity, or lie. This happens in the analysis of different kinds of images, from photographs to paintings and also in the analysis of scientific outcomes that can be in the form of images, diagrams, graphs, or sets of "raw" data. The interpretation of images could be deviant; anyway it gives information revealing features of the interpreter whose subjectivity cannot be neglected. It is important to keep in mind that the interpretation of an image is not to be considered as the essence of manipulation; in fact often the interpretation has multiple shades because different interpreters report different space- and time-dependent points of view. Therefore it is important to discuss images acquired only in response to aimed queries and in the meantime it is useful that the interpreter is conscious that an image is a sign (something that stands for and means something else) (Peters 2017; Eco 2015). Considering images as signs of natural or artificial tracks, it is important not only to know the story of the objects that leave the traces but also to know the questions posed by the investigator prior to the encounter with the facts. Exactly as for Kandinsky, only the environment allows a faithful comprehension of the meaning of the point.

Опыт—Experience

What does one see in a mirror reflected in a mirror?
Michael Ende

The sensations felt and the events lived are elaborated by the human mind and exposed through visual arts, science, and literature. Unknown phenomena are often connected to known ones through analogies so that life, society, and historical context are merged in the reflection of what the artist or the scientist perceives. In the everyday life experience, reflection implies the existence of a mirror: watching one's own image reflected in a mirror is the first human figurative experience, the fictitious representation closest to reality. By analogy if our mind was an ideal mirror it would loyally reflect objects as they are without deforming them, but since perception is subjective, the reflection will be necessarily biased. It could be said that either the mirror does not exist or it does not reflect adequately, but a lot is related to what people expect to see in a mirror (Lenin 2021; Eco 2015; Zeki 2004; Bertamini et al. 2003).

Electron microscopy images share features with photographs and mirror images. Both researchers and photographers try to catch the very essence of the investigated object transposed into impressed plates and digital electronic recording systems. Electron microscopy images, as photographs and also paintings, are a link (though sometimes a barrier/separation) between the operator and the observer, who simultaneously reflects and is reflected. Sometimes, however, observers are restrained from catching hidden information available to the eyes and not to the mind. The scientist behaves like the artist, both creator and observer, and the borders between the roles become breached and blurred so that they can be interchangeable.

The relation between perception and physical world is a well known source of ambiguity, as it happens in a process rare in optics and often labeled as an optical illusion that can occur in scanning electron microscopy: the mirrorless reflection. As seen in Chap. 2, this process, also known as mirror effect, has been reported by microscopists since 1970 and is still of difficult interpretation. The mirror effect consists of a reflection in the absence of a material mirror; as a result an image of the interior of the microscope chamber appears instead of an image of the sample surface under investigation as expected. Mirror imaging and sample imaging anyway are not mutually exclusive; in fact the result can be also an unavoidable overlap of two images: the one generated via the mirror effect (showing a part of the chamber) and the one of the sample; the weight (percentage) of the mixing is dictated by the properties of the sample, the electromagnetic properties of the beam, the experimental setup, and the historical sequence of the operation (Milani et al. 2010; Belhaj et al. 2000; Luk'yanov et al. 1974).

The image obtained in the presence of a mirror effect depends both on time (related to the dynamics of surface charge distribution of the sample) and on the modalities of reading and re-reading of the observer (number and time interval of the observations).

In its whole an electron microscope could be considered as both the source and the detector of the signal (electron beam) which explores the world, reminding the old Aristotelian mechanisms at the base of human vision. This old model is not in accordance neither with modern optical physiology nor with the actual characteristics of an electron microscope where source and detectors (SE detector and BSE detector) are non-coincident, differently oriented, and spatially separated.

The mirror effect shows that image and reality are separate manifestations of the *praxis*; often univocal correspondence between the two does not exist and the analogy between eye and any type of camera or any kind of reflection process (even paintings) is not always completely satisfying.

The mirror effect is the climax of the various issues that can occur during the interpretation of an electron microscopy image (nature of the sample and its preparation, beam generation, electron interaction with the sample, amplification and electron-photon-electron conversion, discretization, electron collection). Trying to make clarity in the interpretation process, it could be useful to investigate, and in this case the semiotics approach is a useful tool, the similarities between a mirror image and a picture or a painting with particular attention to the truth, i.e., the relation between percept and real object. In any case an impressor temporally or materially different (in the case of mirror present and transient, in the case of photographs and paintings past and relatively long lasting, thus constituting a source of memory) is present. The fundamental difference is that the impressed plate or the painting is a material track (imprint, sign), whereas the mirror does not leave any traces (Fig. 6).

To better comprehend images it is necessary to recur to an additional and unusual instrument as literature and art. Science can be seen as sources of images and metaphors and hence offer the rigor to elaborate thoughts which in literature will become new images, in an eternal cycle of mixed traces, signs, and information belonging to both the past and the present.

The complexities suggested by electron mirroring find correspondence in two Magritte's paintings that open the question of what people expect to see in a mirror (Bertamini et al. 2003). In *The False Mirror* (1929) the artist forces normality into an impossible image by placing a cloudy sky inside the eye pupil; in *Not to Be Reproduced* (1937) the painter gives an example of a false mirror, relevant in

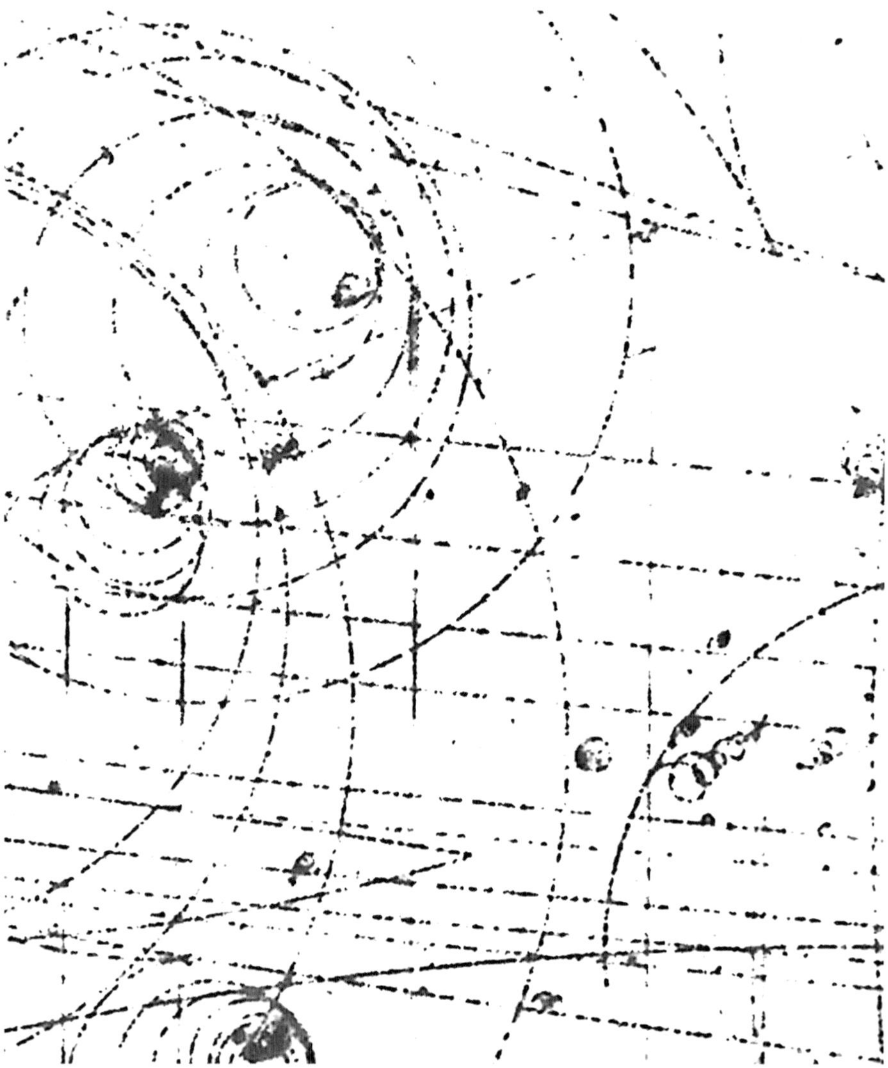

Fig. 6 Traces from the first example of production and decay of Ω^- particle in a bubble chamber (Mathieu and Rossi 1979). In the field of visual products shared by science and art, it can be interesting to keep in mind Kandinsky's "Diagram 25—Line—Linear structure of the picture Little Dream in Red" (1925) (Kandinsky 1947). The suggestive characteristic of this comparison lies in the fact that Kandinsky adds a narration to the visual impression. The artist associates straight and curved lines to different types of forces; similarly in the bubble chamber the shape of particle trajectories is influenced by electromagnetic forces acting on positive and negative charges

psychological debates, highlighting the discontinuity of the world and the alteration of the subject's relation with space and time. This psychological side of the man-mirror relation debouches in the physical realization of mirror images represented in visual arts by self-portraits, and in literature by the internal turmoil that characters face when they look at their real selves in the mirror.

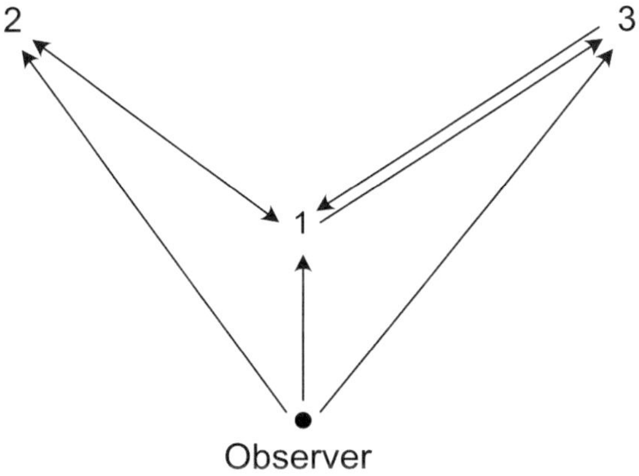

Fig. 7 Sketch from Johannes Gumpp's *Self-Portrait with Mirror and Easel*. The painting was created in two versions: the first, *Double Self-Portrait at the Mirror* (1646), is round shaped and visible in Florence at the Uffizi Gallery; the second, *Self-Portrait with Mirror and Easel* (1646) is rectangular and is part of the private collection of Mr. Peter Mühlbauer (Germany). It is a triple self-portrait: the artist portrays himself while he is painting (1), his face reflected in the mirror (2), and his face portrayed on the canvas (3)

One of the first works in which the relationship between the author and the public (observer) came into play is Gumpp's *Self-Portrait with Mirror and Easel* (1646), where the subject (painter) and the object (painting) contribute almost simultaneously to the perception of the observer. The portrait communicates with the observer who becomes at different levels another subject, i.e., the external observer who compares the mirror image and the portrait. The mirror appears in "the open circle between the seeing body and the body that is seen," enabling the subject to observe himself as another, thus providing an external observer with the recreation of his image as seen in the mirror reflection which corresponds to how effectively the painter sees himself (Cederboum 2009) (Fig. 7).

The personality of each one has inside three Johns: the real John, John's ideal of himself, and John as seen by the others (Holmes 1858). Pirandello developed this thought stating that the real John does not exist; hence he is nobody. John's ideal of himself is one and the ideals of John as seen by the others are a hundred thousand (Pirandello 1992; Oliveri 1959). The multiplicity describes the process where the central self changes and splits into different selves, in a range of other identities, roles, and appearances.

In Gumpp' *Self-Portrait with Mirror and Easel* the encounter of literature and art is evident and suggests a correlation with Eco's *intentiones* classification (Eco 1984). In the painting the author is depicted as: Johannes as he is, represented by the portrait in the looking glass (*intentio operis*); Johannes as he believes to be, represented by the painted portrait (*intentio auctoris*); and Johannes as the others believe him to be, represented with his shoulders turned to the observers, without revealing his identity (*intentio lectoris*). The same architecture (based on multiplicity) is also present in the first scene of the movie *Bridge of Spies* (2015), where again the author's awareness of the existence of an observer leads to different representations of the subject: a spy with a double identity. The famous sentence by Oscar F. O. W. Wilde "There is something fatal about a portrait. It has a life of its own." could be used to describe both the seventeenth-century painting and the modern movie (Wilde 2019).

Images have a temporal development; they are never definitive and always susceptible to further clarification and broader approximations (Lecourt 1973). Hence the image, though connecting both the painter's and the observer's worlds, giving information about their collaboration and competition, remains an external entity that shows more than what could be said in words (Black 1964).

The encounter of artist and mirror is essential and of primary importance. A famous example is given by Michelangelo Pistoletto's *Quadri specchianti* (mirror paintings) where the artist fixes an image on a mirroring surface and then the observer, when looking in the mirror, completes the picture. The observer naturally becomes part of the picture together with the reflections of the world that are incorporated in and by the picture, giving to the relation between image and background further importance. According to Pistoletto the viewer has to inhabit the depicted space that is characterized by a changeable structure which is renewed each time the installation is moved or an object or a new spectator is placed nearby (Pistoletto 2017).

Literature and art show how the mirror is considered the tool of knowledge, and underline the necessity of a philosophical category of reflection. The cognitive value of art and literature finds a basis in the materialistic theory of reflection, where the reflection is seen as an active appropriation of the external world by the thought (Di Marco 1976; Lecourt 1973).

Reflection obviously presupposes the existence of mirrors; indeed reflection and mirrors are inextricably linked even if reflections without mirrors exist (as for instance four-wave mixing or self-focusing in the nonlinear optics of laser and electron microscopy) (Garmire, 1966; Askar'yan, 1962). The mirror has always been used by thinkers and philosophers to clarify the nature of things and as a tool to achieve knowledge.

In the picture by Gumpp the artist retouching the portrait underlines the intention of the painter to affirm the superiority of the painting over other means of reproduction of reality, mirrors included (Baptist 2000). Can real mirrors lie? The conflict between true (or faithful) representations of reality and falseness, together with the role of approximation, is evident. The following table, which extension can be consulted in the Appendix, represents the complexity of the answers that can be given to the question in different fields of education and knowledge.

Table 1 shows that the mirror must be considered a richer tool than a simple flat, reflecting surface; moreover mirrors represent different degrees of complexity so that sometimes they become immaterial.

Table 1 "Do mirrors lie?" The table collects some of the answers and considerations with which scholars, artists, and scientists tried to explain their point of view about mirrors and reflected images. The presence of radically different conceptions of mirrors in theory of reflection is evident both in the two cultures and in the same field

"Un jour deux miroirs s'étant rencontrés s'arrêtèrent pour un brin de causette l'un en face de l'autre. – Tu vois quelque chose? demande le premier. – Non, rien du tout, dit l'autre. – Moi non plus."	Tournier M
"This way [of creating a poem] is not difficult. It can be quickly managed anywhere on earth - most quickly if you are prepared to carry a mirror with you wherever you go."	Plato
"Optics seems to know a lot about mirrors."	Eco U
"There is something deeply disconcerting about mirrors … Despite our constant use of mirrors, our nervous systems remain surprisingly ill-equipped to grasp the mechanics of refraction and reflection."	Martinez-Conde S
"Thomas Herriot and Johannes Kepler discussed the phenomenon of partial reflection and partial transmission of light in diaphanous bodies. The question was, how could an apparently continuous body transmit and reflect at the same time?"	Meinel C
"Before the mirror." ""Cubic metre of infinity" (is) set inside a reflecting cube. Entering the cube, the visitor can only comprehend with his intellect the infinite reflection without any images in the mirrors that look solely at each other, while he can see with his eyes the reflection of his own self multiplied from one mirror to the other on the walls of the reflecting room."	Falciani C
"People have a naive understanding of how mirrors behave … Our commonsense conceptual and - to a lesser extent - perceptual understanding of reflection is limited and biased."	Hecht H
"Is an object reflected in a mirror perceived differently from an object that is seen directly? ... Is the unreality of the looking-glass world reflected in the way that we interact with pictures that do and do not contain reflected objects? … Mirrors produce some perceptual distortions … The difficulty we have in the understanding of mirror reflections … People are undisturbed by paintings with physically impossible depictions of people looking at themselves in mirrors … People discount mirror information even when the mirror provides information not otherwise present in the scene."	Sareen P
"Jeune fille devant un miroir." The girl reaches out to her different selves. "Mirrors multiply personality … and I mean here multiplication is a very positive self as a claim of growth and development."	Picasso P
"A piece of glass is a terrible monster of complexity."	Feynman RP

Artists used and knew different kinds of mirror, one of most used was the convex one. The convex mirror appears for instance in *The Arnolfini Portrait* (1434) by Jan van Eyck. Because of their association with light, mirrors had numerous mystic connotations throughout history, and traditionally reflected images have been valued for their revelatory power. The convex mirror in *Three Ages of Woman* (1510) by Hans Baldung, known as Grien, has prophetic and demonic characteristics. Given the widespread use of other mirrors in paintings and engraving, the convex mirror is rarely encountered in paintings after 1550 with few exceptions as Caravaggio's *Martha and Mary Magdalene* (1598) (Janson 1985).

All sort of things in this world behave like mirrors.
Jacques Lacan

Entering optics, many worlds open, such as quantum electrodynamics, quantum optics, and quantum electronics; these worlds can be visited taking Feynman as a guide to explore the nature of light and the process of reflection. Far from being a mere change in the direction of a light ray, the reflection of light is an event that goes against all the conceptions of common sense.

For reflection to occur, it is essential to consider light. Often overlooked, light is actually the generator of the image together with the mirror and the object. As already highlighted in the comparison between electron microscopy and optics, in the case of electron microscopy, it is obvious and unavoidable to take into account that the image generator, together with the sample, is the primary electron beam. Often the topic of reflection of light is popularized in an apparent simplicity that actually the descriptions are not loyal to the phenomenon, and the reality/truth is greatly distorted (Feynman 2006).

The analogy between an intuitive and superficial "physical optics" and the theory of reflection justifies our interest to analyze in detail classical optics (at the basis of reflection) and discuss properties and behaviors typical of less familiar branches of optics (such as quantum and nonlinear optics where feedback and self-effects have a role). Actually, the knowledge of optics as well as of physics, even at the university level, is usually confined to the realm of ray optics plus some references to wave optics effects, such as interference and diffraction. A confirmation can be derived from the optics *syllabi* of the majority of physics courses for medical doctors and biologists. Some exceptions can be found—but this derives from specific personal educational paths.

From the comparison, both new and hidden features embedded in the theory of reflection can emerge when discussing the interaction process between subject (atoms and molecules in optics) and object (the environment in optics). In optics the theory of reflection becomes materialistic, involving the interplay of energy and matter, and even the subject is material (atoms and molecules). Therefore, optics turns in a good guide to the exploration of knowledge, allowing the identification of characters and their connections. The acquired knowledge spans towards psychological and philosophical issues as highlighted in the following table (Tab. 2).

As previously seen distortions can occur, hence the accuracy and objectivity of reflections need to be questioned. Reflection, from late Latin *re-flectere* "to bend back, bend backwards, turn away," is associated to the mind with the meaning of "turning of the thought back upon past experiences or ideas, remark made after turning back one's thought on some subject" [https://www.etymonline.com/word/reflection]. Reflection is at the basis of knowledge and, due to the interaction between the world and the observer, it requires a discussion of its emergence from its buildup.

Following the thought of Leonardo, to discuss the knowledge we recall the painting in which a picture is an entity or a phenomenon in a homomorphic relation with some prior object. Therefore, when looking at a picture we get at least a part of the same information about the object as we would obtain by direct perception, thus getting both a direct and a mediated perceptual knowledge of the object. The basic risk is to mistake a picture for its object, or even to think that all the qualities of pictures are identical with those of the objects they depict (Bertamini et al. 2003; Karpatschof 2000). This exchange finds its origin in the separation, typical of classical physics, between observer and object and is partially overcome by the appearance of quantum physics in 1905, which revolutionized the way of describing matter and its interactions.

It could be said that the mirror gives an icon of the object, i.e., an image with all the properties of the object represented. This consideration connects the theory of reflection to the theme of the double (mirror image or shadow), with cultural and psychological implications.

The generation of a photograph or a picture cannot be separated from the author's thought and memory (Haverkamp 1993); the reading of a painting, a photo, a text, or an image in the mirror involves also the knowledge of the interpreter, even though in natural sciences the tendency is to represent results (in our case electron microscope images), neglecting the cooperation of the human psyche.

A glass sheet with cracks (broken mirror) presents several fractures that do not have the ability to reflect anymore. When observed or photographed, the broken mirror shows the existence of objects that should not appear in the reflection because they are entities devoid of matter that however take the role of real objects (black lines). Therefore a photograph of the fractures is actually "a picture of the vacuum" proving that objects seen in the mirror can be far from reality: in some cases the reflected images of a subject become objects, that sometimes are a double of the real object and the reflections act on the observer as a feedback mechanism (Katvan 1978).

The role of feedback is well illustrated by laser dynamics where at least one of the mirrors of the optical cavity, due to its reflective properties, is a source of feedback that not necessarily implies self-regulation, and sometimes can be source of catastrophe (the destruction of the laser), thus making an experiment impossible even on a conceptual basis. A good example is the installation *Città dell'Arte* of Pistoletto in Biella, Italy, where it is possible to verify in first person the multiple reflections in a system made of two mirrors facing each other. When the mirrors are parallelly disposed (0°), the only reflected image that the mirrors give is the one of the observer (passive system that absorbs energy from the environment). If one or both the mirrors are progressively tilted of few degrees obtaining a misalignment (active system that modifies and amplifies the signal), the mirrors give a series of multiple reflections, all replicas of the observer .

Feedback also has literal examples (for instance *The Thousand and One Nights*, William Shakespeare's *Hamlet*, Virgil's *The Aeneid*, Homer's *Odyssey*) in which passively narrated characters become active narrators with a technique called "*mise en miroir*," the name underlining possible amplifying aspect of the process that unpredictably leads to univocal, multiple, or uncontrollable catastrophic outcomes; the possible catastrophic results justify the use of "*mise en abyme*" as a synonym of "*mise en miroir.*"

In order to better understand the approximations and the limits of the theory of reflection applied to knowledge, in Table 2 we transposed the elements and processes of the theory of reflection in the language of optics for the interaction between matter and electromagnetic radiation. The reasoning leads to the necessity and importance of introducing the so-called self-consistent field approximation (SCFA), i.e., the reduction from many-individual problems to a single-individual

Table 2 Synopsis of mirrors' properties associated to the theory of reflection and optics

Mirror (physical optics)	Knowledge (theory of reflection)	Mirror effect (particle optics)
Mirrors reproduce the visible aspect of an object. The copy is: accessible only through vision, spatially separated from the original, non-permanent, constantly changing, and strictly related to its original via a reversible specular path	The relation projected real is not a specular reflection: there is the optical reflection of the object and the percept of the object The reflection opens up a plurality of interpretations even in those that say that mirrors can lie	The mirror image can reproduce the visible microscopic or macroscopic aspect of an object, usually not belonging to the target of the investigation. The copy is spatially separated from the original, non-permanent, constantly changing, and it is strictly related to its original via the unique electron/ion path, which strongly depends on the environment
Reflections, generated by mirrors only in presence of light, can be partial or total Reflections also occur in the absence of mirrors, when light traveling in a material meets the environment, even the vacuum When the amount of light is reduced, quantum physics is fundamental to study the properties of reflection	"A mirror that does not reflect things correctly could hardly be called a mirror." Does it exist in nature the possibility to generate a reflection without a mirror? Yes, but… "The reflection should not be seen as mirroring." "The mirror of the theory reflects in its own special way."	Mirror images generated by electrons can be partial or total (usually a mix with different weights of the mirrored chamber and a portion of the sample image) Reflections also occur in the absence of mirrors for instance the reflections generated from BSE
Ambiguity		
Conceptual	**Perceptual**	**Conceptual**
The partial reflection from a mirror as well as the coexistence (at times) of two images can be a cause of confusion or ambiguity In fact the reflection may contain the subject (self) and the object (again the self or portions of the external world)	The reflection is approximate, susceptible to distortions, and never final	The partial reflection from a mirror as well as the coexistence (at times) of two images can be a cause of confusion or ambiguity. In fact the reflection may contain the subject (chamber) and the object (the sample)

Table 2 (continued)

When a real mirror is broken, confusion and ambiguity in the reflected image derive from the different organization of the resulting pieces which behave as independent differently space-oriented mirrors	The perfunctory reflection distorts reality, introducing absolute and relative truths	When the charge distribution is not flat, charges behave like a broken mirror that projects confusion and ambiguity in the image, since the total image derives from the overlap of images generated by the different portion of the electric field
Dynamics		
The laser model is a good example to study the multiple aspects and properties of the mirror in the theory of reflection. Generally the simplest description of a laser system requires the presence of one or two mirrors to possibly realize an optical cavity (resonator) for modal selection and feedback enhancement. Laser is an active optical device, whereas usual mirrors are passive ones. Bistable devices are active, whereas a slice of glass that also behaves like a mirror is a passive optical system (Feynman, 2006)		
Nonlinear (quantum) optics	**Perception**	**Nonlinear (particle) optics**
Specularity, as the possibility to invert the photon pathway, verifies only when subject and object are not affected by their connection. When a material undergoes modifications due to its reflected signal, nonlinear effects occur generating an active reflection Specularity is not maintained in nonlinear optics	The reflection is active, in open contradiction to the commonly assumed passivity of the mirror image Subject and object constantly change during the interaction; they evolve mixing aspects deriving both from randomness and feedback "The mirror selects, it does not reflect everything."	The charge distribution (the electron mirror) undergoes modifications due to the incoming and reflected signal Nonlinear effects occur, generating an active reflection. It is an obvious consequence that we are in the presence of nonlinear electron optics and therefore specularity is not maintained
Coherence		
Emergence of coherence and subject-object interaction		
The physical aspects (nonlinearity, feedback, coherence) of the complex interplay between massive (electrons) or massless (photons) particles and the object (the sample) provide a powerful model to discuss the concepts and the mechanisms of reflection in the philosophical domain. The changes driven by a feedback in the total reflectivity of mirrors cause the emergence of different levels of coherence establishing an analogy with the subject's evolution and transformation in the presence of a mirror		

one. This implies that the status of the system (atom or laser) is considered approximately as a combination of single-individual states (electrons). This method takes into account only the main part of the interaction between the individuals, but not the total interactions. Results obtained in optics show the importance of a set of approximations to describe the evolution of the system regardless of its quantum mechanical nature (dipole approximation, slowly varying envelope approximation, rotating wave approximation, plane wave approximation, mean field approximation, adiabatic elimination of fast variables). The laser behavior, based on cooperative and collective processes, leads to the introduction of macroscopic variables for set of subcomponents of the system. As optics helps us to understand the theory of reflection, so the laser behavior helps us to understand how the relation between subject and object works in the theory of reflection, which is the aim of semiotics.

The triple structure of the definition facilitates the overlap between the scheme of the laser and its processes (Fig. 8) and the scheme of semiotics, describing the relations and connections among object/subject and imprint/imprinter (Fig. 9). The definition directly reminds of a system with two mirrors but it makes explicit the

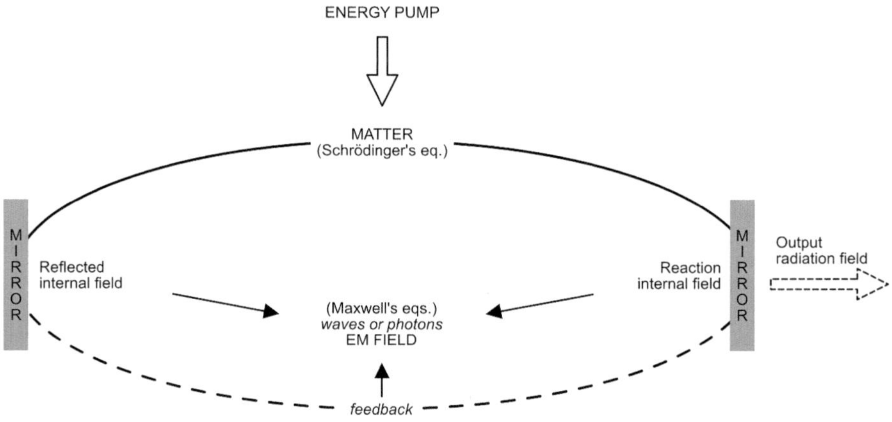

Fig. 8 Scheme of laser activity illustrating the coupling between an electromagnetic field and an atomic system leading to laser action buildup. The dashed part indicates the additional couplings that connect the main actors in the cavity: the incident electromagnetic radiation field (energy pump), the reaction electromagnetic radiation field, atomic polarizations describing how matter is seen by the radiation field, and mirrors (adapted from Arecchi 1966)

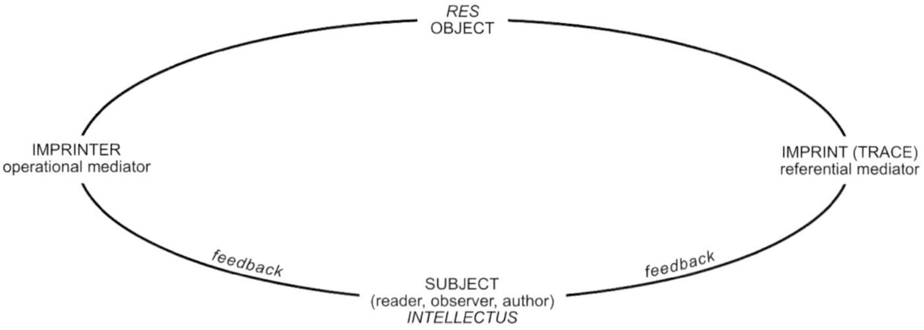

Fig. 9 Representation of the "open circle" or nonlinear oscillator between the seeing body and the body that is seen. In analogy with the scheme of laser activity (Fig. 8) also in this case during interaction the properties of one object are impressed as changes in the internal structure of the other (as reflected in it) and leave in it for some time a trace, an information about the object. A similar scheme has been reported by Jacques Lacan while dealing with the symbolic-real-imaginary realms (Loos 2002). Imprinter or operational mediator includes mirrors and photo cameras

necessity of some underlying approximations, which in the laser give a privileged role to light properties (electromagnetic waves or photon) in comparison with matter (atom), inevitably introducing or wiping out artificial or unrealistic behaviors. Laser behavior is based on feedback (generated by mirrors) and its interaction (interference) with the original signal (the signal at the beginning) and therefore it requires the concept of time. In an evolution towards stability, the two signals are mixed and must be tuned accordingly to their phase difference (another representation of time).

The laser, consisting of an active material (i.e., a material that changes its properties under the effect of light by simple processes like absorption or emission of optical waves or photons), mirrors, and an energy supply, is the canonical example for an optical system characterized by nonlinearity and feedback. Classical and quantum optical nonlinear properties of radiation, together with quantum aspects of matter, nonlinear processes, and feedback which lead the system far from the equilibrium, concur in the laser dynamics introducing stability and coherence cues (emergent properties). This is a prototype of tools useful to investigate a lot of disciplines introducing the concepts of synergetics, nonequilibrium processes and dynamics, open systems, far from equilibrium phase transitions, and dissipative structures (Haken 1978; Fröhlich 1977; Nicolis and Prigogine 1977), and can give interesting suggestions in the in-depth analysis of the theory of reflection.

It should be mentioned that the laser behavior can also be described using cellular automata based on fundamental rules and parallel calculus, but in this work it has been chosen to take into consideration only the serial approach based on differential equations (Costato et al. 1995).

A laser can be seen as a driven, damped nonlinear Van der Pol oscillator, named after the first description in the field of electrical engineering (Van der Pol and Van Der Mark 1927). Modifications of its driving parameters (temperature, current, optical pumping) possibly lead to transient oscillatory dynamics towards a steady state. These oscillations, called "relaxation oscillations," result from the competition between the dynamics of both the electromagnetic field and matter (atoms or molecules). When subjected to an optical feedback (light emitted by the laser is partially reflected and reinjected in the laser cavity) or an optical injection from an external source, these damped oscillations can become undamped and the laser behaves as a self-sustained, autonomous oscillator. As a result bifurcations can lead to more complex dynamical states, including period doubling and intermittency routes to optical chaos (Haken 1970).

As seen so far mirrors are interesting tools to understand the theory of reflection, but an optical system as the laser can be more adapt to comprehend the working operation of the theory of reflection and therefore the approach to knowledge.

The discussion about nanoparticle investigation through electron microscopy entails the themes of comprehension and interpretation of images and texts, that define the aspects of the human experience related to means and processes of communication. From the images we receive a signal, defined as a modulation of a visible field able to attract attention and to transfer messages that require to be deciphered.

The methodology of interpretation of images and texts, concerned with problems that arise when dealing with (meaningful) human actions and their consequences, is part of hermeneutics: from the myth of Hermes, the art and practice of transferring messages with interpretations and explanations (the emissary of gods was "an interpreter, a messenger, wily and deceptive in speech since he was an expert with the use of words") (Plato, ~ 375 BC). Hermeneutics finds its origin and development in Greek antiquity where it was the means to interpret oracles, dreams, and myths; in more recent times hermeneutics handled philosophical and practical works, as well as laws and contracts, implying semiotics, therefore signs and consequently the relation between words and things.

A sign is the constituent of a verbal or visual language carrying a specific meaning. The sign is made up of a signifier and a signified, and in some cases the sign can have a further complexity: the signifier is a complete sign already containing a signifier and a signified, and it is defined as a myth. As seen, myths, sometimes present in science, can distort the meaning of the original sign having a "hidden meaning." Myths intentionally send a message, contain some motivation and intent, and can be considered as second-level signs (Barthes 1957).

In the case of images and texts also symbols are important as tools based on analogues or metaphors used to transmit ideas (sometimes too difficult to articulate in common language) between people of the same culture. Key of this process is the comprehension of texts by the messenger that is fundamental when the message is unclear or ambiguous.

The art of balancing between signs, myths, and symbols in the interpretation of a text, is comparable to the work that needs to be done when interpreting paintings, photographsor electron microscopy images. In the case of electron microscopy images of nanoparticles, the visual message can be incomplete or ambiguous, especially in the absence of the information contained in the data and metadata of the image.

In modern times the interpretative praxis aims to reach the so-called *accessus ad auctores* (access to authors) following seven basic questions, an evolution of the set of questions posed in the Aristotelian Nicomachean Ethics.

who?	*quis*	*persona*
what?	*quid*	*materia*
why?	*cur*	*causa*
how?	*quomodo*	*modus*
when?	*quando*	*tempus*
where?	*ubi*	*locus*
(by) which means?	*quibus facultatibus*	*facultas*

The resulting narration does not produce a perfect, flawless knowledge able to satisfy the ideal scientific method, but it is linked to truth and recalls some keywords that offer a conceptual constellation that builds up experience. The constellation can be summarized saying that the phenomenon of comprehension which "collects all the relations of man and the environment," based on art and poetry, makes it possible "to acquire ideas and to know truth" through personal experiences and practical knowledge (Gadamer 2012).

The oscillator scheme in Fig. 8 can be compared with the one in Fig. 9 that describes human activities that in the simplest form are well represented by the interaction of human beings with mirrors. We can immediately see the presence of feedback and coupling with the exterior which give rise to a much more elaborate and complex structure. Laser action buildup opens the way to a more precise definition (that will be useful in the theory of reflection) of collective processes based on the simple general model of a nonlinear oscillator (that experiences damping or driving forces due to the environment).

The link between the two schemes is helped by the definition of truth (*veritas*) by Saint Thomas Aquinas where truth can be the result of the investigation, and in knowledge the entities *intellectus* and *res* are the counterparts of matter and electromagnetic field in laser (Galatino 2018; Aquinas 1701).

(1) *adaequatio intellectus ad rem* (the correspondence of the intellect to the object)
(2) *adaequatio rei ad intellectum* (the correspondence of the object to the intellect)
(3) *adaequatio intellectus et rei* (the correspondence between the intellect and the object)

"The adequacy of the reflection is expressed in some isomorphism, homomorphism or other type of correspondence between the structure of one interacting body, the reflected, and changes in the structure of the other, the reflecting" (Katvan 1978). Operational and referential mediators are the source and vectors of the message that is exchanged between object (*res*) and subject (*intellectus*). This interaction can be seen as an iterable loop where the exchanged message moves back and forth, suggesting an oscillatory dynamic. Again, the analogy with electromagnetism (represented by the optical case of the mirror-observer interplay) is useful to understand the operations of *intellectus*: electromagnetic waves or photons are the mediators of the interplay between the reflecting subject and reflected object (Fig. 10).

The iterable loop is actually an open circle where the term "open" refers to the necessary interaction with the environment in different active and passive forms (generally objects, facts, ideas that act as driving or damping forces) and, in the specific case, the interaction of an author with the reader. This confirms that in the field of knowledge the laser is a more adequate model than the mirror since in lasers the interaction with the environment has a fundamental role. As shown in Fig. 10 the energy supply from the environment (pump) is necessary for the mediators to deliver the message and therefore to substantiate the object/subject interaction. This is required also for the mirror model that needs energy from the environment to operate, but it appears that the importance of light without which there would be no reflection is for the most part neglected since in mirror operations light transformations are neither present nor considered (Feynman 2006). In fact even though the laser and the mirror models are open systems which require energy from the environment to work, often the mirror model is oversimplified and considered as closed or isolated (the light radiation is taken for granted and its properties remain constant).

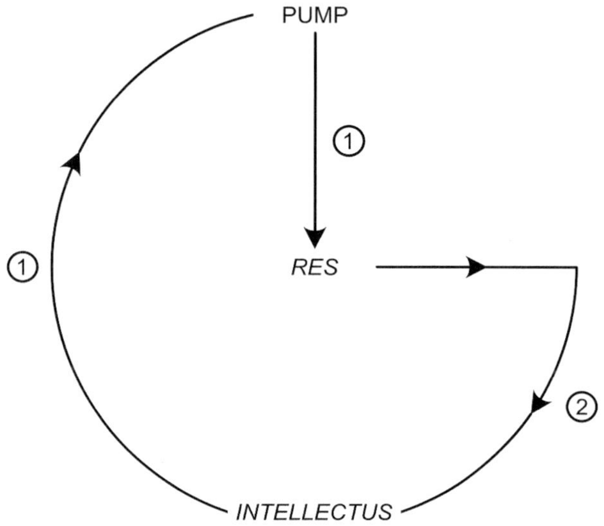

Fig. 10 Structure of the correspondence (loop) between *res* and *intellectus*, seen as an elementary oscillator: (1) *adaequatio intellectus ad rem* and (2) *adaequatio rei ad intellectum*; (1)+(2) *adaequatio intellectus et rei* (equivalent to self-consistent field approximation in the laser model). This elementary oscillator structure is the unit applicable to both *auctor* (author) and *lector* (reader)

The laser is a kind of oscillator similar to a pendulum, a physical system that can be seen as a pivoted rod with a mass attached to the free end and can be studied in several ways: from the viewpoint of elementary and more advanced classical physics, modern chaotic dynamics, and quantum mechanics. In addition, coupled pendula and pendulum analogues are the keys for the investigation of complex physical systems, from electromagnetism to lasers, superconductors, and Bose-Einstein condensates. Moreover, the pendulum has been intimately connected with studies of the earth's motion and timekeeping. It is a very complex system masqueraded as a simple one. The pendulum exhibits a remarkable variety of motions: pendulum dynamics, with and without external forcing and damping, require the essential ideas of linearity and nonlinearity in driven systems, including chaos. Coupled pendula can show synchronization. Even quantum mechanics can be brought to bear on this simple type of oscillator. The pendulum has intriguing connections to superconducting devices like Josephson junctions and biological oscillators as fireflies (Buck and Buck 1968).

The classical pendulum has a long history in physics that was enriched by the appearance of quantum mechanics, although classical physics still provides the imagery. People are getting used to viewing reality at the micro level as sometimes acting like particles and sometimes acting like waves, but one does not know what the microworld objects really are until measured. And even then, we only develop a set of characteristics with which we categorise the material of the experiments, although quantum physics is different from classical physics in some fundamental features. "Under the right circumstances, electrons (particles) can display dynamics that exactly mirror interference effects from wave optics. Similarly radiation often interacts with matter as if the radiation consisted of little bundles (particles) of energy" (Baker and Blackburn 2008).

The pendulum can be used as representation of different types of time as it will be discussed later: the right time when considering the forced harmonic oscillator, the cyclic behavior thanks to oscillations and periodicity, and the damped irreversible behavior leading to the final equilibrium via damping, i.e., energy dissipation. Since it is a universal model, sometimes being the only tool available for systems with (small) oscillations, the harmonic oscillator elicits the interest not only of a lot of scientists but also of philosophers, thinkers, and semiologists as for example Eco (Eco 2007).

Although such systems are abundant in nature, many of them exhibit a self-regulating amplitude which cannot be described by a linear oscillator and require a nonlinear extension; beyond the laser, examples are given by the prey-predator system, by systems generating chemical oscillations or enzymatic cycles, and by biological locomotion and nervous systems. In all these cases the oscillations usually converge towards a well-defined amplitude sustained by nonlinear interactions (stable limit cycle) (Dutta and Cooper 2019; Röhm et al. 2018; Strogatz 2004; Kuramoto 2003; Fröhlich and Kremer 1983; Eigen and Schuster 1978; Haken 1973; Nicolis and Prigogine 1977; Buck and Buck 1968; Winfree 1967).

The mechanism represented in Fig. 10, referred to the relation book/author, can be applied also to the pair book/reader and further pushed to describe the mediation that the *intellectus* can exert between *auctor* and *lector* in the presence of a text or an image (*res*). The result is a more complex system that connects the author and reader and that can be schematized by a couple of interacting loops where the *res* is the crucial point responsible for a strong coupling (Fig. 11). The relation between writer and reader occupied Pushkin's mind: the complex link between the author and reader is based on the concept of inspiration and its duplicity (described by the Russian words "вдохновение—recapture and reconstruction" and "восторг—initial capture," a duplicity that can find a direct representation in a two-level system) (Manukyan 2009).

The laser analogy suggests considering the two loops as two oscillators associated to two single counter propagating modes (waves or photons) of a laser cavity. In classical mechanics the system composed of two oscillators is represented by the compound or double pendulum which can have complex, always repeatable, outcomes resulting in a deterministic behavior illustrated by the Lissajous pictures. Actually the two loops (author/*res* and reader/*res*) are open circles that interact with the environment through a force, a perturbation that in the simplest approach can be described as a further oscillator. Hence the systems composed of three oscillators can lose the deterministic behavior and acquire chaotic characteristics that are strongly dependent on the initial conditions.

The harmonic oscillator in quantum physics is the basic tool to discuss coherence (Glauber 1966). In quantum physics the harmonic oscillator acquires a statistical property that has practical relevance when considering a multitude of identical indistinguishable objects and that "transform" the harmonic oscillator in a boson, a particle able to share with other surrounding bosons the same energy level and that is not subjected to the Pauli's exclusion principle. The reading in terms of quantum

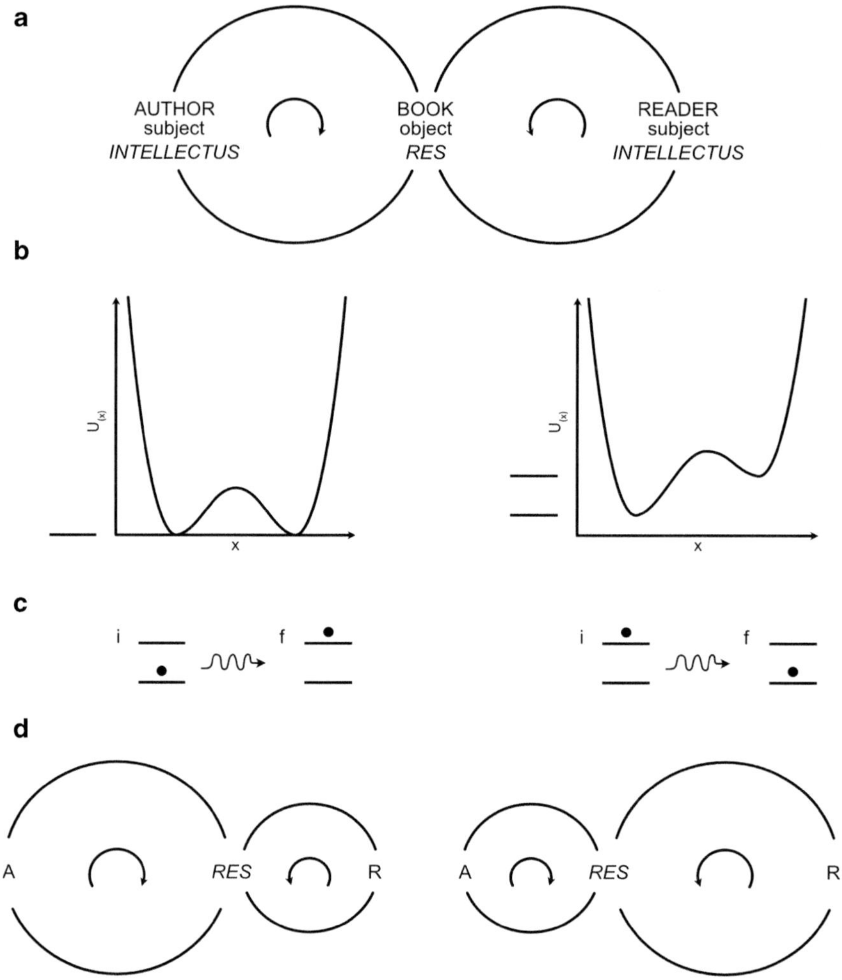

Fig. 11 Two-oscillator system. (**a**) The system, a fermion built up by two bosons, focuses on the relation between the author and reader and can be seen as a switch that shifts the attention on the major role either on the author or the reader, respectively. (**b**) The system in the presence of coupling can be represented by the following symmetric energy diagram that presents the forces at work with a stable state with double degeneracy. When the coupling is strong a symmetry breaking takes place and the stable states are separated; the original system endowed with bistability can be seen as a two-level system. (**c**) The energy diagram allows to investigate the possible dynamics. Considering the system author(*res*)reader, on the left it is represented as the case where the author is given the absolute minimum energy level and the reader the relative minimum (metastable) level; on the right it is represented as the case where the author is given the relative minimum energy level and the reader the absolute minimum (metastable) level. The transition from the initial (i) to the final (f) state in the case on the left can take place only if there is an adequate "energy supply" (that in our representation can derive only from the reader), whereas in the case on the right the transition frees "energy" from the author available to the reader. (**d**) The situation presented in (**c**) can be illustrated by the loop representation where the difference in the circles' area represents the different "energy" associated to author (A) and reader (R), and the subsequent considerations apply to the author as well as to the reader

mechanics of the author(*res*)reader dynamics leads to a population of bosons (immersed in the common field of *intellectus*) that organized in interacting couples (author(*res*; *res*)reader), give rise to a population of fermions: new two-level entities with half integer spin and different statistical properties, subjected to the Pauli's exclusion principle. This passage from couples of bosons to fermions is commonly referred to as "Schwinger representation of a group" (Chaturvedi et al. 2006; Di Rienzo et al. 1978). In the case of strong coupling between the bosons, the two separated levels of the resulting fermion well represent the features of the author-reader relationship and the possible dominance of the author on the reader and vice versa. The evolution of this dynamics leads to a population of fermions composed of pairs of bosons generated by the encounter of the same author with different readers.

Oscillators can approximate a large variety of systems; coupled oscillators allow to deal with systems more complex than those described by a single oscillator. An example of a timed automata system for the description of a double oscillator can be found in Bartocci et al. with a clear description of protagonists and rules (Bartocci et al. 2009). Often coupled oscillators possess an intrinsic symmetry, all oscillators being identical. In this case symmetry-breaking bifurcations, sometimes also called "spontaneous symmetry breaking," can create collective dynamics that no longer reflect the underlying structure of the system. Among others, coupled oscillators can be shown to exhibit chaotic motion, amplitude and oscillation death, coherence resonance, cluster states, and chimera states (self-organized spatiotemporal patterns of coexisting coherence and incoherence in the presence of synchronous and asynchronous clusters of oscillators). The existence of chimera states has been (numerically) demonstrated when the synchronization of two coupled identical oscillators is achieved, even for low coupling strength. The increase of the coupling strenght, together with nonlinearities, can lead to a symmetry breaking that in turn naturally leads to bistability (Hizanidis et al. 2019; Röhm et al. 2018).

A bistable device, especially in the optical domain, provides a good model for the detailed description and handling/control of this problem. Having in mind lasers it is possible to say that this optical system is typically a cavity made of two or more mirrors which have the role to totally or partially transmit the light and also to mix part of the output and input light, yielding a feedback mechanism. Optical bistability is a variant of the general laser model: in a simple approach an active material is present in the laser, whereas in the optical bistability the material has a passive role. Such general models have been extended to different domains (from physics to chemistry, from biology to sociology) and deal with nonequilibrium phase transitions and order of the system, identifying routes from order to disorder, turbulence, and chaos.

The cavity where the light travels back and forth contains nonlinear materials whose indices of refraction respond nonlinearly at light (i.e., I_T is not simply given by I_I times a constant). When two different transmission steady states are available for the same input light intensity, the system exhibits optical bistability, represented by a multivalued function I_T vs. I_I. Optical (or electrical) bistable devices are able to perform optically (or electrically) controlled memory and switching operations. The behavior of bistable systems is expected to be qualitatively different on the basis of classical or quantum mechanics; it can be illustrated by an example based on the harmonic oscillator (considered in the optical domain, where the bistable system is seen as a double potential well, made of two strongly interlaced harmonic oscillator potential wells). In a formal, but easily understandable, language, a single harmonic oscillator can be represented by a parabolic potential well, depicted by a real well with parabolic profiles/walls in the gravitational field; any optical bistable system in the same language can be described by a symmetric or asymmetric double potential well. The main difference between the classical and the quantum approach is the availability of "tunneling processes" (forbidden in classical physics as a consequence of the principle of energy conservation) that establish a new way of communication between the two wells, typical of a bistable situation, thus allowing transitions between the two steady-state solutions.

Again physics provides systems, models, and dynamics useful to treat these ensembles: a very interesting one is the Josephson junction, that being made by two superconductors separated by a very small distance (Fig. 12), gives rise to the

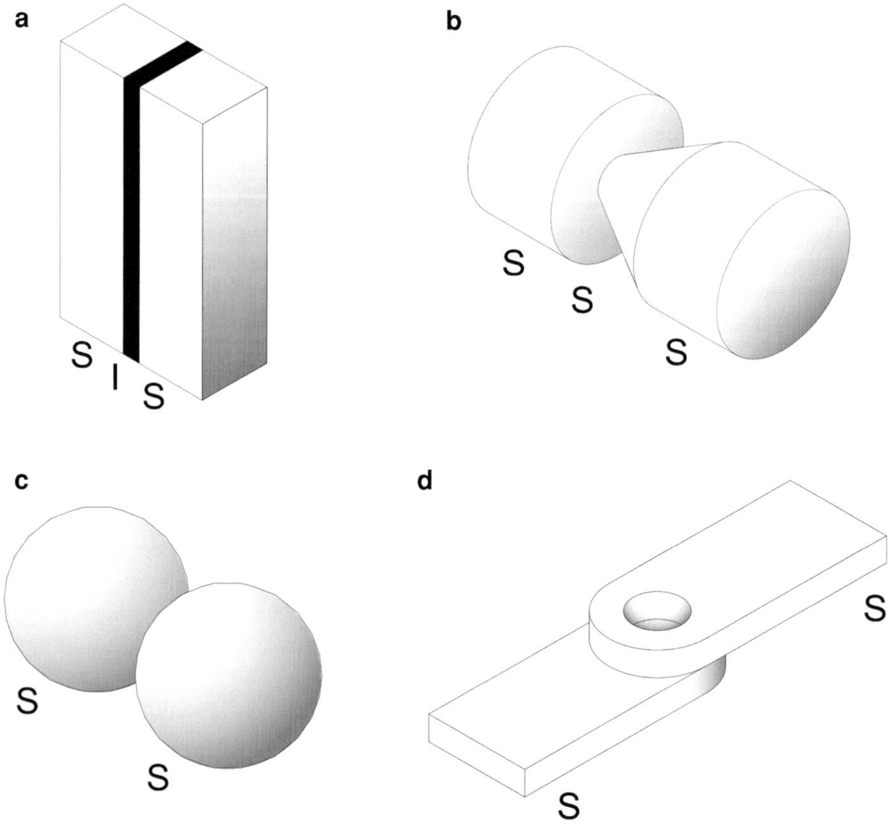

Fig. 12 Different types of Josephson junctions' structures: (**a**) tunnel junction, superconductor-insulator-superconductor (SIS) sandwich; (**b**) point contact weak link (Likharev 1979); (**c**) point contact weak link (Clark 1973); and (**d**) annular junction (Aaroe et al. 2008). (Superconductor = S)

so-called "weak link" junction (Barone and Paterno 1982; Solymar 1972; Josephson 1962). A superconductor is a piece of material where the usual physical entities like electrical resistance and magnetism appear qualitatively different from those in normal conductors. This is due to the formation of the Cooper pairs, structures made of a couple of electrons that, due to the property of the medium to which they belong, i.e., the crystal lattice of the superconductive material, are able to give rise to a long-distance attractive interaction of charges with the same sign. Cooper pairs can flow from one superconductor to another (Josephson tunneling); a phase correlation is established between the two superconductors and the long-range order is "transmitted" across the boundary (Cooper 1956). The whole system of the two superconductors separated by a thin (1 nm) barrier will behave as a single superconductor. Unlike ordinary superconductivity, the phenomenon is often called "weak superconductivity" because of the much lower value of the critical parameter involved (Anderson 2012). Josephson junctions found immediate application in metrology as the basis for all primary voltage standards because they relate fundamental units (volt and second) through a proportionality involving only fundamental constants.

Electronic circuits can be built from Josephson junctions, especially digital logic circuitry promoting applications in the computer domain. Many researchers worked on building ultrafast computers using Josephson logic. IBM started investigating superconducting computers, using Josephson junctions for memories and logic circuit, in 1964 (Juri Matisoo and his colleagues at IBM demonstrated a logic circuit with subnanosecond operation in 1966). At the time, this switching speed was very attractive, for it indicated that Josephson devices could be competitive with semiconductor devices (Mody 2016; Bednorz and Müller 1988).

At a theoretical level, the basic mechanism of the Josephson effects is the quantum tunneling across the junction barrier of the two separate superconductors (considered as two bosons at a macroscopic scale) of Cooper pairs (considered as a different type of microscopic bosons) that are responsible for the superconductors' correlations based on a wide family of weak links different for materials, shapes, and mechanisms. In this way an interaction between two neighboring although separated domains is established. Josephson-like effects could be present in living cells and represent a good model to clarify the interactions between biological systems mediated by electromagnetic fields and suggested by experimental evidence (Costato et al. 1996; Milani et al. 1991; Del Giudice et al. 1989). These devices as coupled oscillators show bistability (Milani et al. 1979) and their couples, SQUIDs (superconducting quantum interference device), can be used as magnetometers to measure extremely subtle magnetic fields for fundamental measurements in biomagnetism: in neuroimaging to map the cerebral activity associated to the magnetic fields produced by the encephalon (magnetoencephalography), and in magnetic resonance imaging (Sprawls 2000; Clark 1973). Arrays of Josephson junctions illustrate interesting aspects about correlations and coherence of the elements of the array and show both the possibility of maintenance of coherence in a suitably prepared array (superradiant behavior) and the appearance of self-synchronization (phase locking) in the array dynamics, starting from a completely uncorrelated initial condition (superfluorescence). Coherence and incoherence of elements of an array of Josephson junctions can coexist in the so-called chimera states (Hizanidis et al. 2019). Both behaviors are associated to the creation of strong correlations among the phases of the junctions during time evolution (Bonifacio et al. 1984). Zimmer's data suggest that a large number of Josephson weak links connected by superconducting paths could be coupled together coherently, therefore reintroducing synchronization mechanisms (Zimmer 1967). Synchronization is a cooperative phenomenon, a temporal analogue of phase transition encountered in statistical physics, and a beautiful, analytically tractable model (Winfree 1967). The most successful attempt to model distributed synchronization has been proposed by Yoshiki Kuramoto (a detailed description can be found in Acebrón et al. and it shares a lot of features and results with quantum optic models, lasers, and bistable optical systems) (Acebrón et al. 2005). There are two basic paradigms for coherence of microscopic oscillators: Bose-Einstein condensation (responsible for superfluidity, superconductivity, and condensation in dilute atomic gases) and lasing (Eastham et al. 2003). These results were possible applying quantum optical techniques to superconducting tunneling junctions that reconduct again to the importance of the laser as a fundamental model for the study of collective properties and self-organization, as emerges for instance from the study of collective dynamics of laser arrays (Previdi and Milani 1998). Moreover the choice to use the laser model to describe the Josephson's junctions as well as the author/*res*/reader correlation has been well-considered since the laser model is general, can be applied to different types of lasers, and can cover a large domain of physical mechanisms and large ranges of sizes (from submillimeters to meters).

As seen so far, optics giving an in-depth characterization of the mirror can suggest how to improve the theory of reflection and how to achieve some insights for the knowledge, avoiding ambiguous traps. When talking about mirroring, beyond the already considered components (subject, object, and the mirror), it is also possible to talk about a mirrorless reflection that requires nonlinear processes as in the case of BSE and SE in electron microscopy (Macherey and Balibar 1978).

According to Dominique Lecourt, the first proposition of Lenin's epistemological theory of reflection is that thought does not reflect an existent reality, and the second is that reflection should not be seen as mirroring (Lecourt 1973). These two considerations are perfectly illustrated by electron microscopy where in some cases the imaged content does not correspond to the real object of investigation, whereas in other cases components with no mirroring properties act like mirrors, so that the images can present distorted reflections as already discussed. Electron microscopy clearly exemplifies the fundamental role played by the mediator of the reflection (electrons in electron microscopy, light radiation in optics); in the case of Lenin's theory it would be stimulating to discuss which is the mediator (the messenger), taking into account that "praxis gives the criterion of the objective truth of knowledge" and that knowledge is always susceptible of approximations so that it never converges to a final status (Lecourt 1973).

In the previous schemes (Figs. 8, 9, and 10) it emerges that a self is always required and the optical analogy can be useful to identify it. One fundamental reflection is indeed the mirror reflection of the subject where the reflected image and the subject merge into two equal selves; the subject recognizes the reflected image, but, due to the reflection, undergoes modifications so that in the end there are two distinct selves (Karpatschof 2000).

The multiplicity of selves can be generated by each element of the theory of reflection. There could be a multiplicity of objects or the presence of multiple personalities (selves) in the subject; moreover the reflection itself can assume characteristics typical of a reflection generated by a mirror broken in multiple pieces. The reflection from a broken mirror is thought to give a reduced knowledge of reality because of the lack of some pieces (Hecht et al. 2005); however the broken pieces can generate multiplicity acting as a set of little misoriented mirrors. This process common in art and literature shows that reality loses its objectivity and that the set of mirrors, each with its relative misorientation, gives a unique reflection of life (Oliveri 1959). The fragmentation of the self is a key topic in various works as for instance in Vladimir V. Nabokov's *The Eye* where the main character's fragmented personality exists only reflected in other people's brain (that acts like chance mirrors) (Nabokov 1992).

The proliferation of the selves triggers a competition between the selves that can lead to tragic consequences as it happens in Wilde's *The Picture of Dorian Gray* (1890), where the character splits into two entities, the forever young Dorian and the aging picture, that seem independent but are actually strictly related (Wilde 2019). The proliferation of the selves is present also in Robert L. Stevenson's *The Strange Case of Dr. Jekyll and Mr. Hyde* (1886) where the two men (Dr. Jekyll and Mr. Hyde) being two faces of the same body eventually converge in one entity (Stevenson 2010).

This section showed that the theory of reflection is useful for investigating the way in which we see and interpret images of nanoparticles. The investigation on nanoparticles requires as first thing to abandon common concepts, renouncing to any convenient visual representation of objects. Since it is not possible to talk about the correctness of images, with the concrete possibility of creating frustrations in those who need to see to believe, the fall of the concept of "representability" through images, illustrations, or schemes is fundamental for the acquirement of knowledge on nanoparticles. Besides, the necessary process of translating images incorporates ambiguities associated to conceptual and perceptual knowledge that human intelligence is able to overcome thanks to the implicit belief that mirrors, as pictures, contain more information than they actually do.

Cognitum est in cognoscente per modum cognoscentis.
The known must be in the knower after the manner of the knower.
Saint Thomas Aquinas

Икона—Mental and Visual Image

Wheeler proposed that repeated acts of observation give rise to the reality that we observe, but offered no detailed mechanism for this.
Brian D. Josephson

Many scholars share the ancient Aristotelian belief that one cannot think without the picture, a view that can be defined as dichotomic or Manichaean. Plato overcomes this dichotomy introducing a dynamic dialectics parting the ideal from the sensible, which interact reciprocally. The characteristics of the concept forming in the subject (eidos) are opposed to the object (eidolon) but still connected by the process of the vision; it is not a case that the same Greek root is at the basis of both eidos and eidolon concurring in the processes of vision and recalling the actions of seeing and looking (Cassirer 2009). Seeing is a passive process that aims only to copy; looking at means to take in all the image giving to details the necessary time and amount of attention performing "an intellectual act of configuration" (Cassirer 2019).

Moreover images, seen as copies or reflections of reality, having a symbolic nature associated to analogues and metaphors, are not univocal and the knowledge necessary to interpret them requires for eidos and eidolon, and consequently seeing and looking at, to be connected. In the case of electron microscopy images of nanoparticles, the necessity to go beyond the sensation of knowing what a nanoparticle is becomes evident, renouncing to any visual representation and taking into account methods and technical indications proper to an electron microscopy investigation (Wiggers et al. 2018).

This way of thinking is perfectly represented in a *pièce* by Bertolt Brecht on Galileo where the main character says "They may be vapors, they may be spots, but before we assume they are spots – which is what would suite us best – we should assume that they are fried fishes. ... We shall go ... at a snail's pace ... and whatever we wish to find we shall regard, once found, with particular mistrust" (Brecht 1960).

The reasoning of the Brechtian Galileo finds its roots in Plato's considerations that associate stars, intended as points in the sky, to the everyday experience of embroideries, highlighting the mind's ability to shift the attention from the reality of objects (set of points) to the abstractness of their connections seen as straight or curved lines. Also in the *Allegory of the Cave* shadows are interpreted as real objects notwithstanding their abstractness and ambiguity; this is exactly the case under investigation in this book, since nanoparticles in a TEM image are the visual representation at the same time of both spots (dots with different shapes and sizes) on a screen and shadows derived from the shielding that real objects exert on the impinging electron beam.

The relation between continuity and discontinuity in the real world introduces the concept of infinite associated to the iteration of the process of division, contrary to what Aristotle said about mathematicians "...they do not need the infinite and do not use it" (Barnes 1991).

Obviously the nanoparticle is a divisible entity, and it is not possible to divide it infinitely because it is not continuous. What attributes meaning to nanoparticles is the context that on one hand separates the interior from the exterior, and on the other hand this separation also represents a link between interior and exterior. In art the context, the shape, is represented by either the frame or the line that contributes to characterize the point and define its properties (Kandinsky 1947). The context is helpful in the observation process, but sometimes it is also an element of disturbance, as for instance in the case of noise. The context defines the extension of the space and helps to isolate the substantial elements on the background from the accidental ones. This skill is at the base of the scientific method with its ability in identifying and dismissing higher-order effects. In other words, this is the basic feature of a Galilean physical approach to nature: to simplify the problem as much as possible, looking for the fundamental characteristics of the phenomenon and neglecting the "details" that sometimes could open to new unexpected science domains and problems.

Another characteristic useful to classify particles is the size which is helpful to distinguish the nano- and the microparticles and becomes fundamental when handling mixtures of large nanoparticles, macromolecules and small microparticles. It would be necessary to understand and characterize the role and the properties of that small amount of matter that should be added to a nanoparticle to turn it into a microparticle or subtracted to a microparticle to turn it into a nanoparticle, especially in relation to emergent properties, dissipative structures, or patterns that could appear (Anderson 2012; Fröhlich and Kremer 1983; Eigen and Schuster 1978; Haken 1978; Nicolis and Prigogine 1977).

The passage from micro- to nanoparticles is a good bench test for the investigation of the partitioning of bulk matter into nanoparticles which implies the relation between classical and quantum mechanics. According to classical dynamics, large particles (portions of matter) can split into smaller ones, and under the proper circumstances those smaller particles could reunite into larger ones, thus introducing on one hand reductionism and on the other hand complex systems theory (Caldirola 1974).

The reduction of the physical phenomena to a finite set of fundamental equations is an approach that generates loops, as it appears for instance in Einstein equations that lead to infinite models. This method has been successful for the construction of the fundamental interaction theories at the base of all observed phenomena. The whole universe anyway is not always the simple sum of its parts; in fact the reductionist idea that to explore the overall behavior of a phenomenon it is sufficient to break it into its fundamental parts is not always valid (Barrow 2012; Dyson 1996). The individual components of a physical system may interact with a strength ranging from a weak level to a strong one and, when two causes yield an effect that cannot be described as the convolution of two separate effects, we are in presence of a nonlinear behavior. An example of this is represented by the property changes of a spring, intuitive visual representation of a pendulum or harmonic oscillator; when under an excessive stress the spring is deformed, reversible or irreversible reshaping, that refer to either the deviation from the original shape or to the realization of a new one can occur. This process is general and characterizes human activity so that it is common not only in a scientific domain but also in literature, for example, when reading and re-reading a text, or in artistic paintings where the support's physical borders, with their geometry and extent, give to even a single point different meanings. It is important to recall that both a text and a picture carry a message that is read within the surrounding context and with the stock of signs and pauses available (Calvino 2009; Haverkamp 1993; Kandinsky 1947).

The operation of reading a text relates to the reader, text, and author. The nonlinear features of this established relation are well described in Figs. 9 and 10. A text as well as a photograph operates on the reader's mind by rooting or concealing themselves in the folds of memory, to the point of influencing the readers. Therefore, the loop resulting from the iteration of reading the same text or picture modifies the state of the reader's memory; this is true especially in the case of photography which is mainly devoted to the memory storage strictly dependent on the material support susceptible to changes more or less fatal due to the passing of time. Reading and re-reading the same image implies the iterated comparison between two mental images: the first associated to the first reading of the photograph, the second generated from the second reading of the same photograph. The comparison is a process that establishes a before and an after and therefore introduces the concept of time that in its reversible form is at the base of physics but in its irreversible form is at the base of life, leaving to the physics of irreversibility the task of rational bridging the two forms (Prigogine 1973). As previously mentioned, the reader, being in a strict relation with the environment (even the historical environment, thanks to memory), is not an isolated entity but always susceptible to relentless changing. The remark of Calvino that a classic is a book that has never finished saying what it has to say perfectly fits to any photographs where "feedback" and time-dependent "memory" are the key players of the action of reading (Calvino 2009). Reading and re-reading activate the reader's memory and evolve in a second-order process that involves that the reader as a subject faces the reader as an object, as it happens when a person self-reflects in a mirror.

This situation is well represented by the question "What does a cell see?" that introduces other than the experimenter role the "independent" role of the cell that sees a world of its own, even though the results of the observation are led by the initial questions of the observer (the researcher) (Lynch et al. 2009). The two views mix in a kind of stereo observation where the experimenter and the object of the experiment (the cell) equally contribute to the research. Actually this mechanism is biased since the view of the cell is embedded in the experimenter's view, recalling the features of a matryoshka doll.

In the investigation of the interactions of bacterial cells, polyurethane, and polyurethane nanoparticles, electron microscopy images represent a means on one hand to understand the cell's knowledge of its constantly evolving surroundings and on the other hand to describe the processes that the cell activates consequently. These processes imply features not directly visible that are ruled by nonmaterial components of the matter, aspects commonly neglected as the electromagnetic field which plays a key role not only at the nano- and micro-level, but also at the scale of universe (Barrow 2012). When handling non-engineered nanoparticles the neglected electromagnetic aspects have fundamental manifestations in biodestruction, aggregation, and adhesion processes, in the uptake and release of nanoparticles, and in the dynamics of the protein corona. In our case nanoparticles derive from a biodestructive process of bulk polyurethane that is commonly thought of as a *continuum*, but in the light of electron microscopy images it is better to consider the plastic material as a complex and discontinuous system made of elementary units. Components of a complex system (subsystems) are usually considered independent and linearly coupled; this limitation sometimes leads to catastrophic consequences on knowledge preventing the comprehension of the behavior of nonlinear systems.

Simple accumulation or summation of objects, data, properties, and concepts, though forming the logical foundations of classical physics, is far from representing the behavior of nature. This approach implies that the initial state univocally determines the evolution of the system and that strict determinism and reversibility hold throughout. On the contrary probabilism and irreversibility are typical of quantum mechanics where the observer must face with the multiplicity of the available solutions embedded in the wave function and hence plays the role of chooser of the possible options (Tarasov 1980). Dealing with collective systems implies the reconciliation between the micro- and macro-world where sometimes the definition of "fundamental unit" (for instance atom, microparticle, nanoparticle, bulk) is arbitrary. The contradiction between the microscopic and macroscopic description poses a set of philosophical problems as the reconciliation of the idealized version of the human mind, which investigates nature based on time-reversible equations, with the biological basis that implies an "arrow of time" (Prigogine 1981).

Reversibility and determinism that characterize classical physics imply the introduction of chaos, passage well described by J. Henri Poincaré in his quotes.

> *A very small cause that escapes our notice determines a considerable effect that we cannot fail to see, and then we say that that effect is due to chance.*
>
> *It may happen that small differences in the initial conditions produce very great ones in the final phenomena. A small error in the former will produce an enormous error in the latter. Prediction becomes impossible, and we have the fortuitous phenomenon.*
>
> *Henri Poincaré*

Poincaré refers to fortuitous phenomena (the representation of chance) deriving from very small causes, side elements, or elements considered marginal by the researcher that associated to errors and mistakes suggest the same mix of terms in Pushkin's verses.

Since in classical mechanics necessity dominates, the element of chance and consequently the introduction of statistical laws must appear when considering aggregates of objects or assemblies of particles, whereas in the behavior of an isolated object, the Laplacian determinism excludes the chance.

On the contrary necessity and chance are present in quantum mechanics, as it is well illustrated by the fact that it is impossible to indicate precisely time and space coordinates of the return of an excited microsystem to its ground state without any external influence, being such a return a random act. In addition to elements of chance, Schrödinger's equation, fundamental in quantum mechanics, provides elements of necessity (Tarasov 1980).

Modern physics recognizes the observed object as strictly dependent from the observer, so that the absence of any external influence is required. This does not imply a return to metaphysics, but it indicates the limitations of quantitative observations. Actually knowledge about the observed system is subjected to limitations that were first identified by Werner K. Heisenberg. These limitations linked to uncertainty make the observation of the world an integral process and *a priori* prevent the possibility to know precisely all the variables of a system, establishing yet another distinction between quantum and classical physics. The uncertainty principle (1927) was formulated as follows: "the more precisely the position is determined, the less precisely the momentum is known in this instant, and vice versa."

General arguments point out that a similar uncertainty relation exists also for time and energy, and in 1945 Leonid I. Mandelshtam and Igor E. Tamm gave rigorous mathematical formulations of that relation calling it "the uncertainty relation between energy and time in non-relativistic quantum mechanics" (Arbatsky 2006). It needs to be noted that time does not share the same formal characteristics as energy, position, and momentum.

In quantum mechanics the observation process changes the state of the object, not the object itself; on the other hand, in electron microscopy the iteration of the observations alters the object itself, adding yet another source of uncertainty, representing an example of the "very small causes" cited by Poincaré that could escape our attention.

Determinism is one of the relationships between human beings and the world that relies on the awareness of causality, i.e., the connection of phenomena through which one or more causes under certain conditions are responsible for an effect. This effect can influence either the future, generating new effects or the past going back to the original cause, thus defining the multifaceted mechanism of feedback associated to the concept of time, including its evolution (time lapse) and its direction (Spirkin 1975).

A good example of what is said so far is represented by the photographic process which implies the presence of an object, its photograph (that represents the object at a given time and stores this representation), and the reading and the re-reading of the picture that hence imply a certain time lapse. In this way the time axis, generally seen as a straight rigid thread, becomes shapeable, enriched by loops and knots in a sort of complex stitch knitting that extends in both spatial and temporal dimensions.

The preliminary considerations about time and space require the concept of trajectory, intuitive and common in classical mechanics, not so easily definable when considering the movement of small particles as the electron and its quantum cloud. From the perspective of quantum physics the prediction of an electron's trajectory, according to Erwin Schrödinger, becomes tenuous. The classical theory says exactly where the electron is at each instant of time without giving experimental evidence, the quantum theory introduces probabilities of finding the electron at different points in space as confirmed by experimental measurements (Scott 2012).

The possible agreement between the two theories is fundamentally ruled, but not exclusively, by the dimension of the considered objects, so once again when investigating nanoparticles it is compulsory to define what a nanoparticle is and in particular its size range.

The dismissal of the trajectory and of its link with movement and time was theorized almost in the same years (1920s) by Kandinsky. Since the line is a "point in motion," all the features of the point flow into the line, including form and dimension. Moreover Kandinsky had already declared the non-absoluteness of a structure, since an element like the point is not an absolute, being its nature also dictated by the environment and the observer (Kandinsky 1947). This introduces the mechanism of resonance linked to the appearance of oscillations and waves that emerges for instance when studying the dynamics of a particle (electron) in the presence of a potential well or barrier, i.e., when the particle interacts with an object. Hence Kandinsky highlights the nonlinear mechanisms of painting. The observation of two independent objects originates the representation of a complex system/object that should be the linear combination of the representations of the two original objects, but this is not true: the observer assumes the role of a field (*intellectus*) in which the two objects are immersed.

Of course, rewinding all these considerations makes the concept and the operational definition of time much more complicated than it is assumed to be in classic or quantum mechanics (i.e. as used in Newton or Schrödinger equations). Quantum mechanics requires the use of probability and probability amplitude making it is necessary to discuss the rules of probability addition associated to the nature (incompatibility) of events. In the probability theory used in classical physics, it is always implied that events are distinguishable (Tarasov 1980), but in reality it is necessary to consider not only the incompatibility but also the distinguishability of the events. This is the novelty in the passage from classical to quantum physics as well as one of the contributions of Kandinsky to the theory of painting.

The rejection of the classical individualization of an object is quite fundamental. In classical mechanics objects are known to have individuality, since in principle it is always possible to enumerate them and observe the behavior of any one of them. In this case, however alike two classical objects may be, they are never identical and can always be distinguished; on the contrary in quantum mechanics two microparticles of the same type should be treated as absolutely identical (Tarasov 1980).

This is a consequence of the Heisenberg uncertainty principle. The generalization of experimental data brings us to a basic postulate, the indistinguishability of identical particles (fermions and bosons) that in turn leads to the Pauli principle: in a system of identical fermions it is impossible for two or more particles to be in the same state. Of course one can only apply the Pauli principle to systems of weakly interacting particles, where one can speak about the states of separate particles, even if only approximately (Davydov 1976).

The novels and tales by Nabokov, Dostoevskij, Pirandello, and Wilde provide a literary representation of the multiplicity of identical systems (with possible different non-observable individual characteristics) through the main characters that are both fragmented and unitary, just like a mirror that is broken (Wilde 2019; Dostoevskij 2009; Nabokov 1992; Pirandello 1992).

The exclusion principle is at the base of the structure and the dynamics of the universe. Bosons, governed by the symmetry principle, are allowed to aggregate in a single coherent state as seen in lasers, superconductors, and superfluids. The same happens with the different fragmentations of the experimental results of observations that, at a higher level, melt into a unique entity, a kind of archetype with the role of bridge between sensorial perceptions (images) and ideas at the base of the genesis and development of scientific concepts (Pauli 2006).

The distinguishability of elements (events) is a classical (macroscopic) problem that in the past brought to the problem of the identification of indiscernibles and, more recently, to the aspects of identity/similarity in a set of pictures or in relation to the represented object. Hence the distinguishability of elements is linked to the relation between identity and concept and to differences between classical and quantum mechanics (Black 1964). Literature is a good guide to see how the macroscopic world seems decidedly unruly and intractable to a classical dynamicist. The analogies that could lead to metaphors, paradoxes, similes, or synonyms between physical models and literary images are not incidental, but rather inevitable.

Interesting correlations between quantum mechanics and literature were made in relation to Oblomov, the protagonist in the world of Ivan A. Gončarov, and Cinderella, the famous fairy tale by Charles Perrault (Perrault 2011; Gončarov 2006; Tarasov 1980). Oblomov has a vain aspiration to do something great, but he is characterized by complete inertia and apathy towards everything that takes place in the world (Dobroljubov 1859). Oblomov, who potentially could do everything but whose fancy wishes have no visible effects, plays the same role as the wave function in quantum mechanics: both Oblomov and the function contain all the possibilities, but the answers are retrievable only by an observer, external to the system, and nonetheless strongly interacting with it. Further intriguing complications could derive from the consideration of the possible existence of hidden variable (Böhm, 1952). Tarasov's considerations are very suggestive and illustrate very well the relation at times weak and fleeting between the observer and observed, responsible for instilling different information in the observer (Tarasov 1980). Further examples of the role of the observer's consciousness of the observer-observed interaction can be found in different artistic forms: in literature with Smurov, the main character of Nabokov's *The Eye*, which exists and does not exist; with Goljadkin (senior and junior), the protagonist of Dostoevskij's *The Double*; or with Vitangelo Moscarda of Pirandello's *One, No One and One Hundred*, and in pictorial arts with the different portraits in Gumpp's paintings, each generated by the painter (Dostoevskij 2009; Nabokov 1992; Pirandello 1992). As the others, even the novel Oblomov provides a tool for the understanding of the principle of superposition of states, and the explanation of the problem of destruction of superposition in the act of measurement (observation).

The Cinderella story should illustrate in a phantastic way the idea of the virtual transitions or an explanation of quantum jumps. This correspondence can further be extended from literature to music, another form of art, following the connections between Perrault and Joseph M. Ravel, and to the psychological analysis of music that presents a strict link with the analysis (and interpretation) of images driven by the brain. Another example is represented by the musical explosions as discussed by Kandinsky: "I saw all the colors in my mind, they stood before my eyes. Wild, almost crazy lines were sketched in front of me" (Kandinsky 2005). According to Kandinsky the interaction of different arts increases their expressive potential and as a consequence also the observer's involvement (Pino n.d.).

Fairy tales, though oft untrue, teach good lad a thing or two.
Alexander S. Pushkin

In conclusion, we can say that the study of nanoparticles requires on one hand the analysis of the identity of nanoparticles as objects and on the other hand the identity of the images of nanoparticles; hence it is necessary to consider the differences between a classical and a quantum description. These distinctions are often based on metaphors that, though useful, sometimes results in ambiguous expressions or representations, and it is known that ambiguity and science are not befriended.

Ошибка—Mistakes

When you think to know what awaits you, you don't see anything anymore.
Steve McCurry

When looking at a photograph or at a painting people see only what they imagine finding. This clarifies the intrinsic structure of imagination which is composed on one hand of biased questions and assumptions and on the other hand of the ambiguity of which organ, the brain or the sensorial system, poses the questions. Surely the information acquired neglecting the senses leads to deceit, and every time the reason judges with total certitude, it withdraws from the truth.

Being realism a recurrent theme in painting, a common feeling is that a "good" photograph is also a realistic one. Artists' approach seems to be oriented towards a more and more faithful representation, although according to some painters and literates this issue is still open. Quotes like "My business is to paint not what I know, but what I see" by Joseph M. W. Turner and "I paint forms as I think, not as I see" by Pablo Picasso show how controversial this topic is, and demonstrate that knowledge, which is based on experience, has a key role in the interpretation of images, as recalled by Pushkin who attributes to experience the role of "the son of painful errors" (Durand n.d.).

The only pursuit of truth is through doubts, hence through questions; actually the consciousness of being ignorant towards a certain topic sets off a multitude of questions and is at the base of research. Often experiments fail, but errors and mistakes can be seen as results as well, that lead to new questions and as a consequence to new uncertainties. Of course, the key to the pursuit of truth is not based on asking thousands of (groundless) questions, but on choosing to ask the good ones (Firestein 2012).

The aforementioned concepts are the reason why this last chapter is built in sections titled according to the questions hidden into the verses of Pushkin presented at the beginning of the chapter (Tab.3). Tab. 3: Questions hidden into the verses of Pushkin presented at the beginning of the chapter synthetized in form of keywords and related to their synonyms.

Table 3 Synopsis of mirrors' properties associated to the theory of reflection and optics

Question	Keyword	Synonym
What is an image?	Discovery	
How does the image represent? (Part I) What does the image tell us about the represented object?	Education	Wisdom, culture, knowledge
What does the image represent?	Experience	
How does the image represent? (Part II) What does the image tell us about the observer or about who generated the image if different from the observer?	Education/knowledge	Wisdom, culture, education
How does the picture lie?	Mistake	Error, ambiguity, illusions, genius, paradox
How is time hidden in a picture?	Time	Case, chance, God, inventor, memory

The previous questions imply a more articulated structure that generates a cascade of questions: How is an image read? What does the photograph transmit? How is a picture generated, perceived, and chosen? In which order is a picture read? How many times is the same picture read? These questions constitute a stairway to a more general key point that highlights the fundamental role of the relation between a picture and its reader, pointing out and at the same time mixing the entangled role of the reader of the picture and the photographer. What is a good picture and how does one know it? One pragmatic reply has been given by the famous photographer Steve McCurry who, on phenomenological and operational bases, affirmed that a picture is good when it says something (McCurry 2012).

The photographer builds an image choosing his own point of view and the right time. The interpretation of such an image is subjective, dictated by the multiple observers' standpoints interacting with the operator's point of view, definitely setting aside the research of univocity in favor of a multiplicity of interpretations. Thus the criticality of making the "right" interpretation arises. In our case this is important when facing the problem of nanoparticle identification in electron microscopy images.

Often it is thought that integrating a photo together with its negative could help to extract more information, as well as looking at a picture/painting under different light wavelengths (ranging from infrared to ultraviolet and X-rays); in some way it would mean to know a bit more of the story of the picture that contains also part of the history of the author (could we term these elements metadata?). This also happens in musicology where the reading of the score (sheet music) could help to better appreciate the execution. This practice is somewhat hard in the case of electron microscopy images where the concept of negative loses nearly all its meanings.

The building of knowledge is a fundamental function of the brain that is not just a passive narrator of external events, but actively participates to the building of what we see, attributing a meaning to the several signals it receives. In this context the ambiguity is also present, a characteristic that the brain develops with a protective function, where the *stimulus* is stable and the percept varies. A famous example of ambiguity related to science and images is given by the Kanizsa's cube, by the Italian psychologist Gaetano Kanizsa who dedicated his studies to images and their perceptions. One of the aims of the brain is to find a solution to problems such as the reading of images. Where a unique solution is not present and none of the choices prevail, the only option available for the brain is to produce or accept diverse interpretations in the shortest time (e.g., of texts or artworks), each one valid, giving rise to a high level of ambiguity (Zeki 2010).

When the observation has two or more solutions one faces an undecidable problem and cannot decide among the interpretations: this process for humans is a well-controlled introspective experiment, considering the jumping back and forth among the different perceptions that cannot be simultaneous, an activity that cannot be shared with a machine. This is linked to the famous Turing problem where a machine keeps oscillating without temporal limits, unless a decisional mechanism is inserted by the programmer (no matter how refined, it could be just a temporal cutoff). Hence it is important to find the prerequisites necessary to overcome the vagueness associated to image interpretation (in our specific case the identification of nanoparticles), such as mistakes, lies, misinterpretations, illusions, and ambiguities, providing tools for the identification of unknown objects.

Knowledge and ambiguity are commonly seen as antipodes so that whatever is not endowed with a univocal meaning becomes the object of critics and refusal; ambiguity opens to a plurality of interpretations, increasing meanings, metaphors, and allusions. Due to the innate assumption that knowledge must satisfy the criteria according to which truth and false are opposites, and that whatever has not a perfect definition is far from being rational, ambiguity is often given a negative connotation and thought of as an obstacle. Rationality, shared by both science and humanities, seems to bring different outcomes: although the representation of the world is the common aim, science acts through theories and experimentations, whereas art's tendency is to generate ambiguities, suggesting to the observer new interpretative paths to explore. This contraposition can be overcome, for example, through Kandinsky's research about the relation between arts (painting and music) that in parallel applies to science and humanities (Kandinsky 2005). In science the compresence of various interpretations, each one successful (true or false, based on cultural, economical and political currents and changes in the popular thinking) at determined times, is a recurrent element. Examples are the conflict between the Galilean and the Ptolemaic model of the solar system, the discussion regarding the existence of the aether, the wave-corpuscle dualism, and the cancerogenicity of diverse substances.

Among the positive connotations of ambiguity, surely there is the capability to highlight the boundaries (eventually crossable) of the world as perceived by different observers (Lumer and Zeki 2011). Thus the dynamics of the interplay between the observer and the system under observation or between the different observers originate, and the strict relation between observer and observed forcibly results destructive or at least alterative. This is important especially in the problem of measurement in modern physics (quantum mechanics) that is exemplified by the wave/

corpuscle dualism where the properties of the measure instrument mix with those of the observed object and the aim of the observer.

Help to frame the problem is brought once again by literature where the ambiguity of interpretation can be enhanced by the absence of punctuation. An example is the famous Latin sentence of the thirteenth century "*ibis redibis non morieris in bello*" from the Chronicon by Albericus Trium Fontium, an ancient prophecy that, due to the intentional lack of commas, could be interpreted in two opposite ways according to where the pause falls: either "*ibis, redibis, non morieris in bello*" "You will go, come back, will not die in war" or "*Ibis, redibis non, morieris in bello*" "You will go, you will not come back, you will die in war."

Examples of ambiguity have been analyzed by literature in connection with the theme of the double, a physical entity that has all the characteristics of the main physical subject and as its counterpart possesses the same pertinent traits necessary to obtain an *abstract* character. Obviously the presence of the double has to be recognized, declared, and shared by all the members of the society; this praxis, possible in the literary confines, is mandatory in the scientific field being the main criterion for the progress of knowledge (Nilsen 1998). The birth of a kind of proximity effect between literature and science, where science cannot express itself without a language and a narration, is represented as a primary ambiguity embedding a subtler one: the double of the primary ambiguity can be either a doppelgänger (the two characters are indistinguishable) or *rectius* a double (a single character with two distinct personalities). A couple of photons emitted by the same atomic source are a good physical example of this complex structure of ambiguity: they can be seen as two identical units, whereas improving the observation they could have different propagation direction or spins (polarization) and be considered as two faces of the same coin. The emission of photons from an excited atom, the case to which we are referring, is actually more complex: in the phenomenon of the stimulated emission, it is possible to observe the simultaneous emission of two perfectly identical photons, synchronous with the same direction of propagation and the same polarization. This subtle distinction recalls the attention on the points of view: the one of the reader/experimenter that knows the story and the one of the characters/photons in the story that can or cannot distinguish the main character(s). This classification is so fine that Dostoevskij's *The Double* notwithstanding the title belongs, according to Nilsen, to the doppelgänger's category (Dostoevskij 2009; Nilsen 1998). Dostoevskij's work perfectly exemplifies the importance of the standpoints: the reader knows that there are two distinct Goljadkin (senior and junior) and can distinguish them from their behavior, whereas the characters in the story always see just one person and do not recognize that he has two interchangeable personalities.

The ambiguity that arises in literature is amplified when, observing electron microscopy images, objects need to be identified knowing the point of view: the one of the researcher, that of the couple researcher-instrument, and the one of the observer of the final image. Often, in images, different objects correspond to the same spatial set of pixel content, so that they appear similar and are often thought of as doubles in the same way as it happens with Plato's shadows. A problem arises when categories of objects are introduced and the wrong *abstract type* (name) is given to the wrong object, thus opening the issues of identification and falsification.

The visual system is based on a combination of (optical) projection and (interpretative) cognition. The cognitive need of extracting the maximum information from whatever is around us is an existential feature of our nature. It cannot be excluded that in some cases the perception of ambiguous images enters a loop, i.e., an oscillatory state with a characteristic frequency. A useful tool to try to understand perception and vision mechanisms is the Necker's cube named after Louis A. Necker, a Swiss crystallographer and geographer, who in 1832 noted that a two-dimensional representation of a three-dimensional wire frame cube exhibits "a sudden and involuntary change" in its orientation: sometimes one face seems to be in the front and the other face behind as if looking the cube from above, other times the face which was in the front seems to be behind as if looking the cube from below. This object has been used also in psychological studies especially by Kanizsa, so that nowadays the Necker's cube is also referred to as Kanizsa's cube (Mason et al. 1973; Necker 1832).

Looking at the point of intersection between two lines in a Necker's cube, the observer does not know which line is in front and which is behind and visualizes two different cubes or convex trihedra (open solids, made of three lines crossing in a vertex and defining three planes) realizing that the picture is ambiguous and facing a process of perception at least bistable (Fig. 13). The observer has to make a selection between the two different views: the first one is purely casual, probably the more familiar to the viewer, and most likely it will be the view from above. An element necessary to understand this mechanisms needs to be found in the historical/cultural roots of viewers that are more used to seeing things from above (Einhäuser et al. 2004). This is confirmed even when the observer, after watching the ambiguous figure, decides to fix on paper the two percepts. The favorite one will be represented promptly, and only with concentration and more time availability also the other one will emerge and will be reproduced, showing the advantages and disadvantages of speed in reconnaissance operations.

The ambiguity is broken if the amount of information is reduced, paradoxically (Fig. 14). When some of the lines are removed or the focus is on selected parts of the

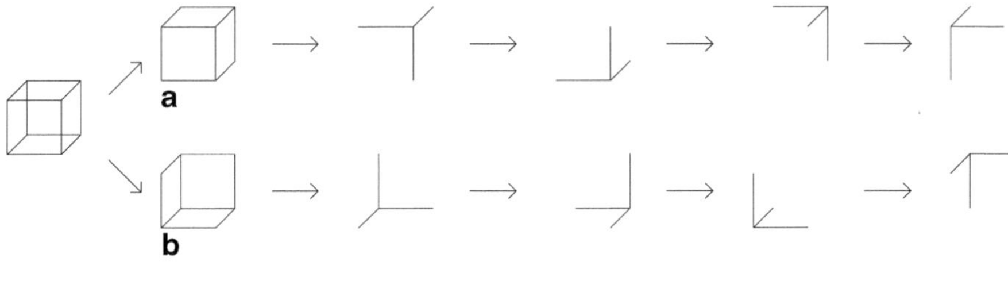

Fig. 13 Views of the Necker's cube. (**a**) View from above and (**b**) view from below. The views are the interpretative cognition of the optical projection (relative spatial orientation) of the viewer and the cube as it is in the observer's mind. Each alternative has a double perception involving convexity/concavity; thus the number of the possible representations increases (multistable perception). For each representation the four trihedra are highlighted

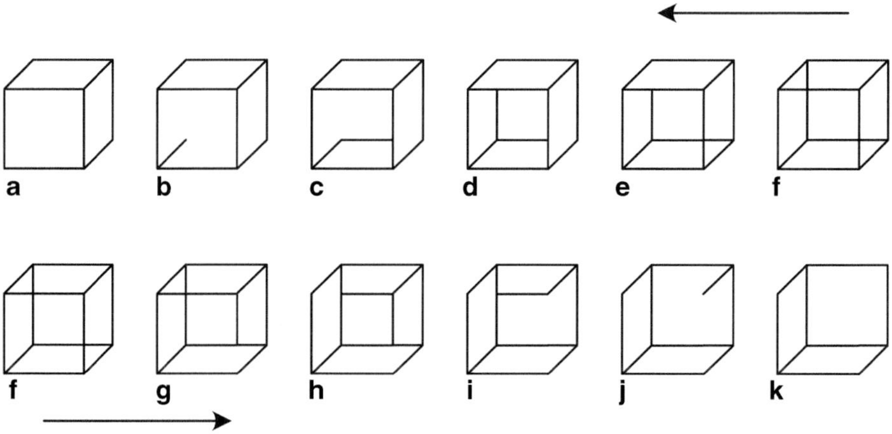

Fig. 14 Possible steps that conduct from the Necker's cube (**f**) to one stable solution (**a**) or another (**k**). The manifestation of the hysteresis is present in the passage from **a** to **k**. Focus on details of the figure modifies the stability ratio between the two percepts (**a** and **k**)

figure, the stability ratio between the two percepts is modified. Moreover when the thickness of the lines is altered, the Necker's cube is transformed in an impossible object.

Another aspect of ambiguity in the picture analysis is associated to the antinomy "concavity/convexity"; usually two-dimensional drawings represent three-dimensional objects, thus creating figures that appear to enter or exit a plane that separates two semispaces. Looking at the Necker's cube the observer has an additional percept of concave objects, intuitively associated to the interior of the cube. More precisely seeing the cube from the outside, from four different points of view, gives four diverse configurations of the figure and ambivalent patterns emerge: the two alternatives of perception have equal strength, usually a particular alternative is preferred and it remains stable until the perception of the other one. The habitude to one percept can imply longer time (on the order of minutes) to reverse the focus on the other alternative. After a sequence of switching, the frequency of the reversion increases until it reaches a stable average value. The cyclic/oscillatory reversion, influenced by will and practice, cannot be prevented and exhibits hysteresis (Brinkman 2010; Brokate and Sprekels 1996; Agarwal and Shenoy 1981).

Focusing on the vertex of each convex trihedron in Fig. 13a, the observer will also perceive two concave trihedra representing the interior of the cube as seen from two different standpoints. The switching time between the percepts (the convex and the concave) can be measured and is approximately constant, in the order of few seconds (Ditzinger and Haken 1989). This optical illusion is now used to test computer models of the human visual system to see if they can arrive at consistent interpretations of the image the same way humans do.

The issue concave/convex is not only associated to geometrical images or to impossible objects, but it has also a fundamental role in the interpretation of images acquired, for example, with an electron microscope or a telescope.

The objects in Fig. 15a, c, though having strongly different sizes (micrometers to kilometers), exhibit a very interesting feature linked to brain/mind interpretation processes. The objects appear to be convex but after a π rotation they appear to be concave as shown in Fig. 15b, d. The objects in Fig. 15a, b are a double that can be seen at times as convex or concave, pointing out the interpretative role of the brain. The same can be said about objects in Fig. 15c, d. The figures show how, according to the direction of the exploring source (in one case the electron beam and in the other the sunlight), the perception of the depth changes from concave to convex; this is also related to our historical/evolutive habitude to be exposed to a light source placed above our eyes. It is important to underline that also the concavity/convexity inversion has no direct links with the actual orientation of the object in the space in relation to the measure system. The importance to recognize this ambiguity in order to avoid illusions and errors when planning experiments that interest the morphology of surfaces is also evident, especially when completely automatic systems are used.

Protagoras asserted that there were two sides to every question, exactly opposite to each other.
Diogenes Laertius

To solve the ambiguity of interpretation it could be useful to learn from how the direction of light influences the perception of concavity/convexity finding partial help in the shadow, a secondary phenomenon associated to the image (Von Fieandt 1938). The electron beam on samples with morphological variations, both on the sample surface or in the slice, due to the spatial asymmetry of both the detector and the sample relative position, redistributes electrons generating the compresence of dark zones and light areas in the correspondence of structures, an effect that recalls shadows. The shadow is an important element that underlines the transition from a two-dimensional to a three-dimensional representation since when "the eyes see a picture they send an image to brain that then has to make sense of" (Becerra and Barnes 2010). Shadows are themselves a form of double associated in a non-univocal manner to the object that generates them, and hence a source of ambiguity usually interpreted by observers through neuropsychologic mechanisms. Shadows suggest interpretations recalling images already present in the memory of the viewer with a link to experienced things, representing a different situation than that of Plato's *Allegory of the Cave* where shadows kept prisoners away from the knowledge of the world. Shadows helping to both underline the convexity or concavity of the structures in the images and to invert perspective, are not always deceitful, sometimes they help to enhance or amplify the knowledge.

In a set of different interpretations, as those in Fig. 15a, b, no one knows which is trustworthy, because a loyal interpretation implies the existence of an absolute reference frame, in contrast with any post-Newtonian physical law. Therefore, in the case of electron microscopy images, what does it mean to rotate a sample when the absolute reference frame is unknown? The problem points out some differences between TEM and SEM imaging operations based on the different orientation and movement of the source of the primary electron beam. When looking at an electron

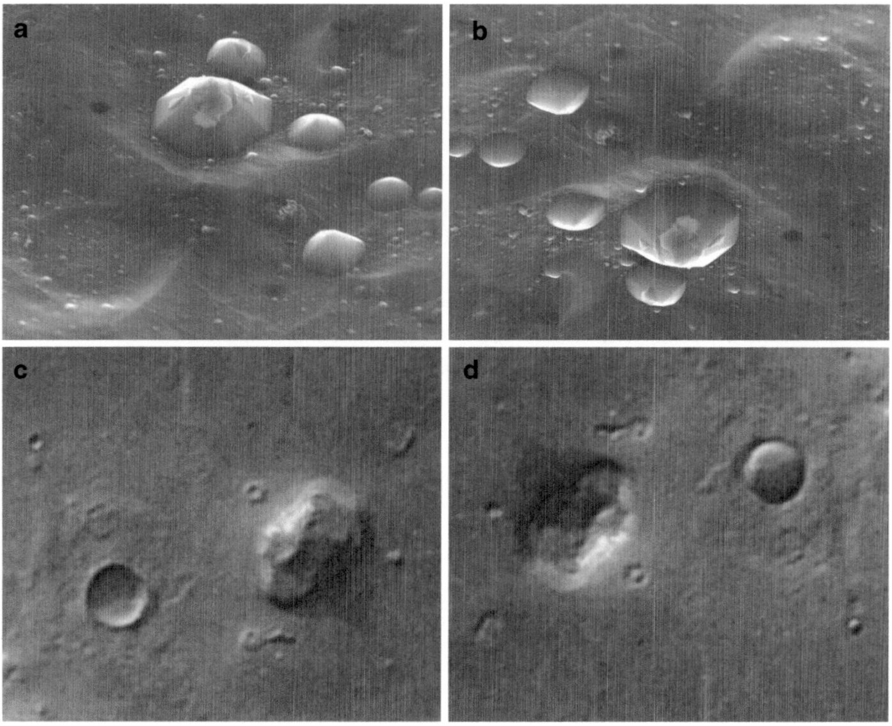

Fig. 15 (**a**) SEM image of a silicon wafer surface after wet etching. (**c**) Planetary imaging photograph of the surface of Mars (Viking, 1975–1976). (**b, d**) Rotation by a π factor in the plane of figures (**a**) and (**c**), respectively. The structures photographed by the Viking probe (**c**) can be compared with the ponds and pyramid-like structures that appear in the SEM image (**a**). The rotation influences the perception of the structures that in (**a**) and (**c**) appear convex, whereas in (**b**) and (**d**) appear concave. The figure shows the necessity of taking into account the mixture between reality, ambiguity, illusion, and mistake in the image analysis. This problem was already faced by the psychologist Kai von Fieandt who posed the attention on the fundamental role of the light source position (Von Fieandt 1938) and in op art (~ 1960) by the painters Victor and Jean-Pierre Vasarely

microscopy image, perception could be altered and sometimes the feeling of handling an ambiguity gets compelling. Even if the object in the image has not been identified, the researcher can still extract information about the building of the image getting back to processes of picture generation. A faulty interpretation or faulty machine, associated to the multiplicity of interpretations in an ambiguous situation, can lead to mistakes and illusions. The ambiguity of the dualism concave/convex can be solved adding another observation system as for instance a system FIB/SEM where the ion beam suitably oriented can proceed to the etching of the expected convex object and the subsequent electron imaging can detect the presence of the etching. Moreover image interpretation is dependent on the analyses of how vision mechanisms work. In everyday life familiar optical phenomena as mirages or Fata Morgana an happen: depending on light, viewing angle and the environment, objects not physically present are seen and the visualization can be only explained by physical laws and used as examples of semiotics' sign and track.

The multiplicity of percepts and the dynamics of perception and cognition is fundamental not only for computer software based on automatic object recognition in images, but also in neuropsychology and neuropathology studies, for example, in relation to Parkinson's disease, which history started in 1817 with the publication of *An Essay on the Shaking Palsy* by the British apothecary James Parkinson (Parkinson 2002). The main topic in these research deals with the correlation between rigidity in motor domain (typical of Parkinson's pathology) and efficiency in non-motor domains, like perception or cognition, when the visual system cannot reach one particular perceptual interpretation and either vacillates (random process) or oscillates (periodic process) among two or more competing partially stable percepts, introducing bistable or multistable dynamics in perception.

The link between vision and perception is composed of some intermediate steps, distinguishable experimentally between voluntary and involuntary processes with characteristic times.

Illusions like the one illustrated by Necker or generated by Necker's cube were already present in ancient times, for instance in Roman or Byzantine mosaics. In more recent times other optical illusions have been created and implemented in graphic arts and a lot of artists inserted these elements in their paintings as, for example, did Escher who painted a lot of impossible objects as in "Belvedere 1959" where he developed the entire picture starting from optical illusions and in particular a "cubelike absurdity" as that of Necker's cube (Koffka 2006; Escher 1972). Moreover Escher points out the interconnected relationship between the observer and the observed (who acts as an observer), showing the importance of the psychological aspect of vision and perception of both "possible" objects and objects that are source of illusions.

Impossible objects are phenomena which cannot exist but which we can see all the same.
Bruno Ernst

The first formal example regarding impossible objects was the Penrose triangle published by Lionel S. and Roger Penrose in 1958 and recreated by Escher in his construction of impossible worlds. "The physical model works from only one special angle," and there is a split between conception and perception of something (Becerra and Barnes n.d.).

The concept of concavity/convexity already discussed and present in illusions, paintings, graphic art, and mosaics can be retrieved also in common life. A simple representation is given by pictures of a drop of water falling into a pond and generating ripples. If an image in the plane xy is rotated 180° upside-down the same ambiguity associated to the presence of shadows appears (Fig. 16). The concavity/convexity issue associated to the physical or digital rotation of the image generates a paradox regarding the fact that the rotation in an xy plane of a figure (a two-dimensional object) can modify the percept of the three-dimensional object pictured along the z axis.

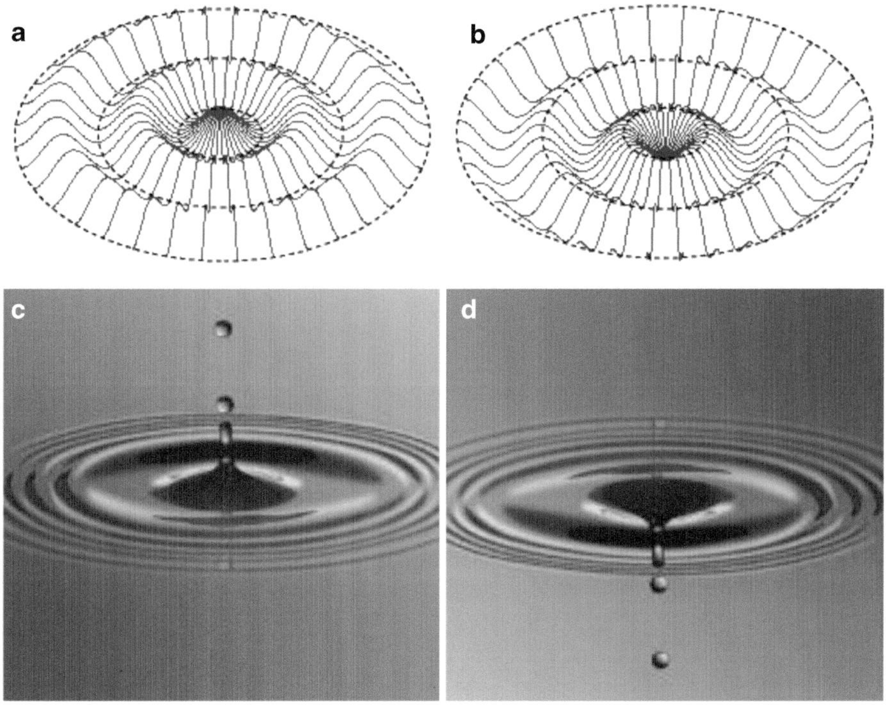

Fig. 16 (**a**) Plot of the process of a wave generation in a liquid; (**c**) real picture of a drop falling into a liquid and generating a wave; and (**b**, **d**) figures (**a**) and (**c**) rotated by 180° in the plane, respectively. The 180° rotation makes evident the punt/dent couple (the punt is the pushed-up area at the bottom of a bottle associated to the dent, intended as the concave part at the bottom of a bottle). Almost everyone sees a punt in (**a**) and a dent in (**b**); the 180° rotation of (**c**) "transforms" the little cone sticking up into the air into a volcano-like dent (**d**). In figures (**a**) and (**b**) the punt/dent transition is not related to any light source, whereas as a consequence of the rotation from (**c**) to (**d**) also the light source is moved as it can be deduced by the shadows

As seen in Fig. 15a, b the perceptive trick of exchanging top and bottom of a structure represented in a plane shows the existence of a problem in the interpretation of electron microscopy images: a plurality of percepts and bifurcations of choices increases the number of possible operations that, if repeated, give rise to an arborescent labyrinth. Percepts generate stories and doubling the perceptions the number of interpretations and in turn of stories also doubles; stories' plots cross themselves, creating a manifold of crossing points, each representing possibilities to switch back, forth, and sideways from a labyrinth to another, multiplying to the near-infinite the generation of stories (Schneider 1980). A retrograde analysis of the arborescent maze thus created allows to go back in time and possibly to identify choices at the basis of new branches. Bifurcations, nodes, or crossing points are a tool to study the appearance of new structures and can lead to chaotic dynamics. This reconducts to Pushkin's verses and to the binomial "chaos and experience," where the two concepts both compete and cooperate: chaos can be seen as one of the possible dynamical outputs of the arborescent labyrinth with bifurcations, and experience can be seen as the reiteration of experiments or observations. The multiplicity of interpretations provokes an unlocking from the environment; given different copies of the same object the observer tries to choose which is the real one, thus obtaining independence and autonomy from the environment because the observer's will, not conditioned from the environment, is autonomous and has freedom of choice.

Experience shows that the two states of ambiguous figures are equally possible and "stable for certain time lapses notwithstanding the instability of stimulus," but they cannot be accepted at the same time (Lumer and Zeki 2011). The observer first visualizes solution A and then solution B, and in order to decide which one is satisfactory he goes back to solution A and then to solution B in a continuous neverending process that sometimes can be chaotic (when the observer's response time is much longer than the oscillation's period) but more frequently oscillatory. Another possibility is given by the fact that the brain, on the basis of previous experiences, is able to choose between two (or more) states stopping oscillations, thus showing that often observation's results are influenced by the background of observers.

This switching is characterized by a reading time of each alternative for the first step of the oscillation (time of permanence in each possible solution plus the transit time between the two) and by a re-reading time (the time spent moving back from B to A, plus the permanence time in the two solutions and the transit time between them). The permanence time in the possible solutions within each cycle can be different in respect to the first reading time. This process, common to a lot of scientific applications, happens also in the brain as proved by medical-psychological evidence concerning the switch (transit time) and lock (time of permanence) process in ambiguous figures reading.

Hence the multiplicity of percepts generated by an ambiguous figure implies the appearance of time. The initial time-independent picture activates in the observer a time-dependent answer in which the brain/mind system creates a split characterized by an oscillating loop and a bistable behavior in the cognitive acquisition (visual perception), doubling the percepted representations (conceptual perception).

Optical bistable devices provide a good model for the description and handling of this problem. Bistability provides hints for understanding processes and for moving from a qualitative to a quantitative analysis of brain activity when deciphering ambiguous figures connected to the transition between left and right well (Fig. 17), each of them being associated to one of the two representations. Moreover, asymmetry in the double potential well (generated by the environment of the system and more commonly by an external agent) can provide a suppression tool for the oscillation of the system between stationary states, leading to convergence, i.e., to state selection (Fig. 18).

The possibility of hysteresis intrinsic in the bistability and typical of first-order phase transition is a hint for "memory" formation (a process related to time) and exploitation in brain, analogously to what happens in physical bistable systems. The question whether the oscillating process converges in a finite time to a univocal result or not is open.

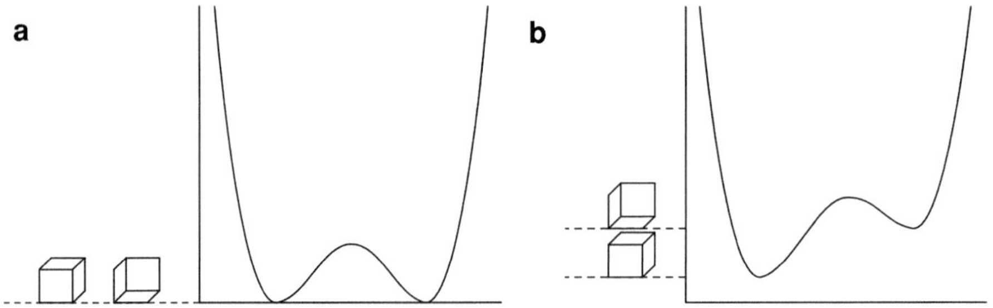

Fig. 17 Symmetric (**a**) and asymmetric (**b**) double potential well. The processes illustrated in Fig. 11b, c are useful to understand the appearance of the two orientations of the Necker's cube as well as the concavity/convexity dualism. The situation in (**a**) represents a degenerate state, i.e., the two visualizations are equally probable, in the absence of the observer. The situation in (**b**) describes the fact that the degeneracy is broken so that there is one preferred, most probable, visualization. The energy potential in (**a**) is transformed in the energy potential in (**b**) by the addition to the original potential of a linear one, representing the constant force due to the presence of the observer

It is well known that in some bistable system there is a difference of results between a classical (macro) and a quantum (micro) approach; this underlines not only the importance of the assumptions at the basis of the models and the evaluation of the real aspects of the system (like continuity or discontinuity), but also the relevant role that the system's interactions have with the environment and the observer (negligible or not). In the classical approach (continuous system) the system is based on the interplay of components progressively smaller without an inferior limit; whereas a quantum approach (quantized system), which seems more similar to the case under investigation, deals with elementary units that can be very small, but discrete (with an inferior limit). Thus we face the dichotomy between a conservative or dissipative vision of the system under investigation.

Once again physics fundamentals are anticipated by writers. "Absolute continuity of motion is not comprehensible to the human mind. Laws of motion of any kind become comprehensible to man only when he examines arbitrarily selected elements of that motion; but at the same time, a large proportion of human error comes from the arbitrary division of continuous motion into discontinuous elements" (Tolstoj 2016). This quote by Lev N. Tolstoj, linked to the reconstruction of historical events, perfectly applies to the boundary between nano- and microparticles, underlying the importance of the premise about continuity and discontinuity in the scientific investigation. Anyway, as far as science is concerned, this sentence looks dated: the appearance of modern physics at the end of the nineteenth century proves that reality, both matter and energy, is associated with discontinuous elementary units, thus privileging a quantum vision that only sometimes coincides with the classical one.

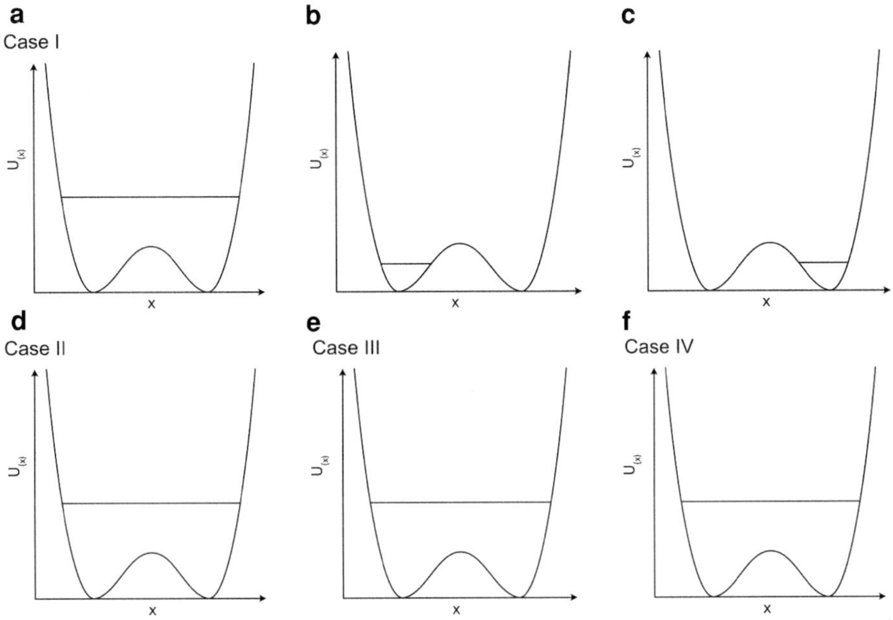

Fig. 18 The tools represented in Figs. 11 and 17 help in the classification of the multiplicity of outputs experienced by the observer when in the presence of an ambiguous situation.
Case I: double symmetrical potential well in a non-dissipative system, classical description. (a) Continuous exchange between the two states. (b) Only state on the left. (c) Only state on the right. The selection among these three possibilities derives from the preparation of the system (initial coupling with the environment that is afterward removed).
Case II: double symmetrical potential well in a non-dissipative system, quantum description. (d) Continuous exchange between the two states due to the tunneling effect.
Case III: double symmetrical potential well in a dissipative system, classical description. (e) Presence of two steady states mutually exclusive either only state on the left or only state on the right.
Case IV: double symmetrical potential well in a dissipative system, quantum description. (f) Presence of the two states visible in an alternate way.
The choice among the three final (long term) possible states derives from the preparation (initial state) of the system (i.e., the initial interaction with the environment-observer):
– Continuous exchange between the two states (P_{left} and P_{right}) associated to periodic stationary dynamics
– Selection of one state (P_{left})
– Selection of one state (P_{right})

		P_{Left}	P_{Right}	
Case I	a	50	50	
	b	100	0	
	c	0	100	
Case II	d	50	50	
Case III	e	50	50	AUT rule for probability
Case IV	f	50	50	VEL rule for probability

As illustrated by Mr. Palomar in "The infinite lawn" discretization can be restrictive; due to the uncountability of the "blades of grass," he states that his meadow is "a collection of grass." The definition that Mr. Palomar gives about the lawn is the key concept; in fact if he would have thought of the lawn as an ensemble of blades of grass, automatically the lawn would have been a collection of numerable objects; since Mr. Palomar sees the meadow as an expanse of grass he explicitly points out the indivisible property of the lawn and makes the collective property of the ensemble of grass emerge. The more Mr. Palomar circumscribes the field of experience, the more it multiplies within itself, opening dizzying perspectives (Calvino 1985).

This discussion (analogy or model for brain activity in ambiguous image recognition) should be completed by considering the case of an asymmetric double well and extending it to the transition from two-dimensional systems to three-dimensional systems. The instability and oscillations typical of ambiguous image reading can be found in the just discussed model of the double potential well that require the introduction of the concept of time; in particular it must be stressed that conservative systems with periodic dynamics oscillate for infinite time (Fig. 18).

In dissipative systems the representative point of the system generally reaches a stable stationary state using also long times (the system can never reach a stable stationary state because of continuous oscillations between the two states). To distinguish between conservative or dissipative systems, it is essential the conservation of real systems with which precise boundaries (but not necessarily stable and static) between the observed object and the rest are built. An operative representation of the conservative/dissipative aspect of a system is given by the dynamics of a computer. The distinction between dynamics of conservative and dissipative systems has fundamental outcomes. In the competition between a computer and human brain, a computer wins if the system is closed as in a play of chess; but when it has to face the dynamics of an open system, the computer loses, even though brain or computer operations do not halt and keep looping forever. The transition between the two states protagonists of the ambiguity could go on indefinitely, thus putting the brain off from operating other functions (Wessen 2016).

It would certainly be nice to know whether or not it is actually going to finish and hasn't just entered an infinite loop.
Ken Wessen

Alan M. Turing studied the problem of incompleteness (that results to be ubiquitous) and demonstrated that there is no way to know if a self-contained computer program (that constitutes a conservative system not necessarily closed) will halt (Anon n.d.) "Why is Turing's halting problem unsolvable?"). Since the brain is asked to operate other functions, it is not a self-contained system, and therefore at some point in the analysis of ambiguous figures it will surely decide to halt the program. Thus if in reality the observer is able to select one of the two competing images, it could be interesting to understand which is the brain mechanism that is activated to dampen the oscillating process.

All that is said about images is based on a strict connection between science and human activities, two spheres that need to be reunited and should communicate between each other (Snow 1959).

> The physical sciences are ... an integral part of our civilization, not only because our ever increasing mastering of the forces of nature has so completely changed the material conditions of life, but also because the study of these sciences has contributed so much to clarify the background of our own experience.
> *Niels Bohr*

Hypothesizing that visual perception depends on visual imagination, focusing attention on the visual aspects of thought is important because it connects what the eye sees and the retinal output with the pictures stored in the brain: visual recognition could not take place without this combination. A bridge exists from the eye to the cognitive science of pattern recognition as demonstrated by the fact that normal experience influences both the anatomy and physiology of the brain as well as the neural mechanisms of perception (Sacks 2010; Hubel and Wiesel 1979).

The role of memory is important for both sculpting the physiology of the brain and for providing a support, a visual archive of the acquired experience. Human memory can be classified in long-term, short-term, and working memory (Cowan 2008). During visual mental imagery, perceptual information is retrieved from long-term memory, resulting in the subjective impression of "seeing with the mind's eye" (Sacks 2010). Since memory is composed of a voluntary and an involuntary portion, often past episodes of the everyday life, triggered, for example, by a sensation or an object, come to life unconsciously. Involuntary memories, the subject of study of cognitive psychology, are also considered as a sort of cognitive realism (Mace 2004). In literature examples of cognitive realism are given by the odor of *lilas blanc* that recalled Wilde's Dorian Gray the "strange period" of his life, or by the madeleines in Marcel Proust's *Remembrance of things past* (or according to other translations *In Search of Lost Time*) that recalled the author's afternoons spent with his aunt (Wilde 2019; Proust 2003; Proust 2000).

Memory is fundamental not only for human beings, but also for computerized systems. Generally the long-term memory does not last forever, because it is subjected to deterioration or to degradation of the support. This is an unavoidable process for humans, but technology seems to have bypassed the problem introducing the memory storage on virtual clouds (Snowden 2019; Von Neumann 2012). This brings back to the problem of reading and re-reading images making comparisons: computers with data permanently saved on clouds have always the same image as term of comparison; humans with images printed on photographic paper or saved on digital devices, or just in one's own mind, have the recall of the image that due to the passage of time, personal experiences, manipulation, diseases, or other biological causes could be deteriorated or biased.

The considerations made so far about the perception of images and the conservation of memory show the importance of the cultural background deriving from both science and humanities. In the period of the Enlightenment science and humanism collaborated without any problem, but in recent years the separation of knowledge between disciplines belonging to the scientific or humanistic area without reference to different goals or method of investigation made the relation between the two cultures problematic.

> *È difficile pensare oggi ad un'interazione così stretta*
> *tra campi del sapere distanti tra loro.*
> *knowledge so distant from each other.*
> *Alessandro Schiesaro*
> (Authors' translation)

As a consequence of the segregation of knowledge the communication among members of the two areas became difficult, finding its peak in the twentieth century with the creation of two cultural systems, seemingly irreconcilable (Schmahl and von Weizsäcker 1998; Snow 1959; Rushmore 1924).

This split was strongly fought by vanguards that promoted and carried on their scientific research in the artistic field (Lyubchenko 2014).

> *The illustrations and tables both artists produced to support their theories are highly detailed and rational, and could easily been mistaken for those explaining phenomena normally scrutinized by hard sciences' lens.*
> *Irina Lyubchenko*

Moreover, the knowledge gained from both science and humanities, merged with the personal experience, contributes to creating culture, intended as the background processes useful to obtain results, hence including all the endeavors and mistakes made.

All that is seen about uncertainty, inexperience, and mistake, also associated to the iteration of processes, requires to talk about time, a concept that can be expressed in two different ways. Time can be seen as a straight line, recalling the arrow of time that represents the inexorable passing of time, or it can be seen as a curved line, recalling the cyclicality of time and the iteration of events.

This section showed the strict link between mistakes and the concept of uncertainty. This correlation was clearly exemplified in the interpretation of ambiguous figures where also time played an important role; in fact situations of uncertainty determine delays in the decisional process, a situation similar to the critical slowing down of the physical processes near a phase transition, i.e., at the point where the system is exactly between two phases.

Uncertainty is also linked to doubt, approximation, and imperfection, once more recalling the concept of error and mistake, all terms with a negative connotation. Actually it is just investing time while recognizing, understanding, and studying our own mistakes that we can walk the way to scientific and intellectual progress, thus attributing a positive connotation to imperfection.

Время—Time

"...si addormentò in un angolo di cuore
per un tempo che non esisteva.
... Fuggì senza allontanarsi,
ritornò senza essere partito...
Così mentre il tempo moriva, restava l'amore."
"...it fell asleep in a corner of the heart
for a time that did not exist.
...It fled without moving away,
returned without having left...
So while time was dying, love was staying."
Antonino Massimo Rugolo (Authors' translation)

Since an image has the fundamental characteristic of blocking time, often photography is defined as a process of "time freezing." The space and time relation between the observer and observed enters in the game revealing that something is observed in some place by somebody that at the time of the photo shoot was not there. Commonly it is thought that a photograph depends only on the time it was taken, but it contains more information about time than expected; in fact the freezing process at least takes a time interval dictated by the exposure time. Consequences are easily visible for instance in blurry photographs that can have at least a triple reading: a mistake, an art work, a technical document (photograph). In the last case the photograph contains information that go beyond the primary aim of the photographer and depend on the reader's skills and aims to decide whether an image is good or not.

Usually people think of photographs as static images developed in a space, but images enclose also time. The observer's awareness that a photograph "speaks" of space and time makes it useful to condense into a single image the dimensions of both time and space. This process led to the development of cartography, a tool useful for the planning of an itinerary or a voyage: in this case the map contains both the time of the past as history/document and the time of the future associated to the duration of the voyage. It can be said that the map, being conceived as a tool useful in a voyage, presupposes a narrative idea (Calvino 1984a, b), as an electron microscopy image of *Staphylococcus aureus* incubated with polyurethane nanoparticles alludes to the itinerary of membrane vesicles and nanoparticles.

The importance of time that emerges from the picture is represented by the exact timing of the picture and by the connection of what exists prior and immediately after the shot. The interaction between the reader of the image and the photograph also implies time because the reading and re-reading of a photograph (installation of the picture in memory; possible variations about the interpretation of the image) are associated to the concept of time and time flow, the last being responsible for the stability or deterioration of records (memory).

Memory and ambiguity together with real absence or voluntary and involuntary cancellation create an intriguing pattern (Benvenuti 2017). Ambiguity in pictures needs time as a cardinal element in the description of the dynamics of oscillation processes between possible interpretations (oscillation time, the time necessary for

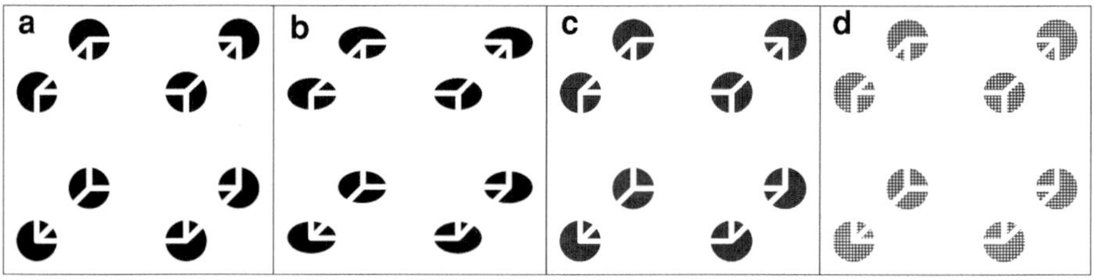

Fig. 19 Influence of the visibility of the Kanizsa's cube related to the shape and structure of the points. (**a**) Original Kanizsa's cube. (**b**) Ellipsoidal points. (**c**) Circular fine-grained meshed points. (**d**) Circular large-grained meshed points

reaching the possible convergence of the oscillation process). The analysis of the reading process of ambiguous pictures introduces the historical and complex question eradicated in the Western thought of dichotomy of time: linear and circular visions, or time's arrow and time's cycle, where the cycle arises from the oscillations between two states and events have no meaning as distinct episodes, whereas the arrow emerges with irreversible sequence of unrepeatable events, when the damped oscillations lead to an univocal reading of the picture (Gould 1987).

A further touch of complexity derives from the presence or the absence of the protagonist, where the absence is as heavy as the presence. As visible in the electron microscopy images of our investigation, the presence of a nanoparticle can give information as well as its absence. The competition between presence and absence finds a thorough illustration in the optical illusion of Kanizsa (1955) where a suitable arranged set of signs creates immaterial, "phantastic" objects (therefore by definition non-objects) that are ambiguous (Fig. 19).

Vision cannot be based on the fact that the eye is like a camera. Seeing does not equal to getting a direct comprehension of reality; on the contrary seeing is like acquiring data from incomplete information. Vision goes beyond the stimulus on the retina being based on the dialectics between the visual system and the direct experience of the world (Cellucci 2015). The action of seeing a photo "produces meanings in us, but at the same time we produce meanings in the stream of images we watch" (Cubitt 2014). This has a strong relation with interpretation and consequently with truth, a concept inevitably linked to mistake, doubt, and uncertainty, since every interpretation and narration is independent on its reliability. It is like living in an n-dimensional world and discovering that others live in a $(n + 1)$-dimensional world; the historical truth is not what happened, but what we assume it happened. In some sense this introduces an unstoppable evolution of reality or facts, as well exemplified by the history of science and by the mechanisms of the re-reading and re-writing that generate interpretations that can be unique depending on the reader's culture and historical context (Hyland 2019).

The Moro Affair by Leonardo Sciascia represents an example of the complex interplay that connects time, time freezing, iteration of reading, observation, memory, ambiguity, reality, and interpretation. The author, while telling the story, reflects about literature and its meaning, in particular the aspect of literature as a tool for the pursuit of truth, of the "real." The writer reasons on a literary adventure that connects himself with three other authors in a mixed cultural game, enveloping the reader in a web; the reader gets trapped and loses the feeling of time and causality. Hence the reader is not only influenced but subjected to the ideas of the writer that deforms so much the story to the point that the reader does not recognize any more the reality or the facts as he believed to know (Sciascia 2014).

Sciascia's work is a good narrative representation of the relation between the observer and image/photograph and of what happens during trials where, having photographs a great value as proofs, also images are considered as tool for the pursuit of truth. In both cases, since each fact is made of minute always subtler events that sometimes become undetectable, images need to be carefully interpreted because they always appear more complex, contradictory, enigmatic, and labyrinthine. The truth Sciascia takes into consideration is not a simple truth but becomes a real truth, as if the adjective accentuated its credibility making it more understandable and reliable; as if the truth exists if and only if associated to the "real" (Belpoliti 2002).

What is discussed so far requires an elucidation about the identificative process that the reader/observer operates. The identification of the false, intended as the absence of truth, is already a step towards the pursuit of truth. A bit as in the Kanizsa's cube the identification of the points of absence makes the truth (the cube) emerge (Fig. 19). These processes are present also in scientific papers and in our case in the research activity about nanoparticles.

Processes of identification of falseness and their classification find help in this activity by literature and in particular by the translation of texts. Translations in Middle Ages were the only texts through which obtain information and they were considered at the same level as the original. This happens also in transcriptions from code to code as well as in the passage from electrons to photons in electron microscopy. As a result we can have:

- Strong False Identification: it is part of each one's everyday life to come into contact with ingenuous or fraudulent cases.
- False Weak Identification or Interchangeability Assumption: it is the case of translations and transcriptions considered unloyal because it is not uncommon that the translator or the writer consciously alters the text, even cutting or censoring it. A process not so much different from the operator activity on an electron microscope in search of the best parameters to capture a picture.
- Pseudoidentification: when identifying an observed set of points with an object never observed in reality. This can happen due to the scientific tradition or to unprecise information about the potential existence of the object in exam.

The dual nature of time, implicit in the relation between observer and image, forces us to consider the reality of time, focusing the discussion on what time really is.

What is time? Who can give that a brief or easy answer?
Who can even form a conception of it to be put into words?
Saint Augustine

It is notable to remember that in the common conception of the causality principle linear time is associated to the irreversibility, but it could also be independent from the irreversibility as it happens in the basic formulation of the physical processes based on the Newton and Schrödinger's equations.

The linear concept of time has had profound effects on Western thought.
Without it, it would be difficult to conceive of the idea of progress, or to speak of cosmic or biological evolution.
Richard Morris

Cyclic time is characterized by repeatability and reversibility. Cyclic time cannot be used to describe a story, because natural processes or historical facts need linear time to be contextualized (Halpern 2018). On the other hand, most people throughout history have held fast to time's cycle, viewing time's arrow as either unintelligible or a source of deepest fear (Eliade 1992; Gould 1987).

The problem of the coexistence of the cyclic and linear concept of time is well exemplified by nature as thermodynamic and evolutionary processes to which cosmology, nuclear decay, and psychology can be added.

The two representations of time, time's arrow and time's cycle, are an example of metaphor, a powerful instrument typical of poetry, literature, art, and science, that creates and transforms, often misleading, and at the same time able to hide or highlight contradictions; metaphors are the circumlocution of one idea in terms of another and help us to understand unknown concepts (ideas) through familiar phenomena (Ross 1989; Cassirer 1946). Metaphors, which cannot be considered as true or false, influence the way of reasoning about complex issues and eventually push people to look for further information, even though people do not recognize metaphors as influential in their decision. A powerful example to investigate straight and circular time is the pendulum, the metaphor by excellence for the scientific method, as well as a simplified model helpful in the organization of thought (Gould 1987).

The arrow-cycle pair can be seen as a dichotomy or antinomy, real or fictitious, of an ambiguous nature, that recalls the dichotomy between wave and particle. This dichotomy, typical in quantum mechanics, is actually fictitious because it compares two situations of a different statistical nature: the single and the multiple.

A biological qualitative and quantitative representation of time's arrow and cycle interplay can be seen from the viewpoint of individuals belonging to a population in a world of two populations connected by the prey-predator relation. In this case a single individual experiences time flowing unidirectionally along a line (birth-reproduction-death); a population, in view of its survival, experiences time as a cyclic turn of events.

Obviously without the time's arrow, time's cycle cannot exist and vice versa: it is not possible or thinkable to disallow time as a line for an individual and simultaneously accept cyclic events in a population.

People don't have a single integrated representation of complex issues..., but rather rely on a patchwork of (sometimes disconnected or inconsistent) representations and can (without realizing it) dynamically shift between them when cued in context.
Paul H. Thibodeau

The metaphor relative to the dualism of time is also a paradox since the dualism of arrow and cycle is not a real one. The geometric consideration helps explaining the paradox since each point of a line and of a circle are in biunivocal correspondence. In the Euclidean geometry there is a transformation that maps all lines to circles, the identity transformation, so that a line is also viewed as a circle (a degenerate circle of infinite radius). This is related to the size of the (finite or infinite) universe in which the individual or the experiment is embedded. Hence linear and cyclic time are eventually coincident; on this basis we can expect that time has no direction.

It is interesting that Kandinsky studied the dualism between straight and curved line according to their subjective description and evocative power. The straight line is intended as the induced form of a unique force working on the point (which moves along the line), whereas the curved line results when two forces work simultaneously (Kandinsky 1947). This feature assumes a relevant role in connection with Newtonian mechanics and in classical or quantum optics.

The dichotomy between arrow and cycle was already present in Ancient Greece where time was a patchwork of representations and was divided in three aspects, each contributing to shaping the experience of time: Chronos, Kairos, and Aion.

Chronos *(associated to the divinity of Saturn)* represents the mere conventional measurement, a means of counting time. This aspect of time is passive and in medieval philosophy is symbolized by a circle (Courtney 2019).

Kairos (associated to the divinity of Uranus), symbolized by a point, represents nonhomogeneous time in which each instant can be more or less worthy. Kairos is dedicated to the subject-situation correlation.

Aion (associated to the divinity of Neptune) represents the infinite time and is symbolized by a square. Aionic time neglects passive or active roles, focusing only on the actions and not on the actors. Aion "the eternal" appears in time (Chronos) as an aperiodic sequence (Kairos).

> *Space by itself, and time by itself, are doomed to fade away*
> *into mere shadows, and only a kind of union of the two*
> *will preserve an independent reality.*
> *Hermann Minkowski*

The circle, point, and square together build the symbol of the salt-point (a dot in a square in a circle) that represents the transcendence of the "self" evolving in the dynamics of the psyche; it has been the object of interest in mathematics and geometry of both Pushkin and Kandinsky (Courtney 2019; Kandinsky 1947). Since antiquity the issue of time has been the object of interest in different fields, from literature to art and physics with Einstein's theory of relativity revealing how time and space are intimately related.

This vision makes indispensable to associate a static reality (as a photograph) to a temporal evolution; in fact every time we look at a photograph we recall that we live embedded in the passage of time (Gould 1987).

In semiotic language images are signs, prints, or impressions that stand for a content and can be possible impressors. Discussing the link between print and imprinter it is necessary to recall that in electron microscopy (as well as in light photography) the link between print and impressor requires at least two steps in which two different imprints, nanoparticles and electrons of the beam, intervene. In the simplest case nanoparticles have their own print on the image, and as a response to the primary electron beam a set of electrons belonging to the nanoparticle (SE) is sent to the detector. Besides, the electrons of the primary beam while forcing the nanoparticles to reveal themselves to the detector sometimes dramatically modify the nanoparticles until they disappear. In this case the imprint on the image is the trace of a nanoparticle no more existent and which reality could be questioned. We can say that the existence of the nanoparticle is no more necessary and all the information are restricted to the image and to its interpretation. In this case the image reader observes the image at time t_1 when the imprint and not the actual nanoparticle was captured. The reader turns into the interpreter of the image assuming that at a previous time t_0 imprint and impressor were spatially contiguous and therefore the concept of absence is *de facto* a temporal absence. This process leads to the acceptance of imprints with no impressor, opening serious problems in the interpretation of electron microscopy images, especially in images where parameters have been imposed: for instance when the black and white filter is applied in the form of a binary code for pixel content reinforcing the contrast (1= white=object from reality; o= black=absence of objects=vacuum). This operation can be misleading since the application of the filter can lead to the wrong classification of the pixel content, attributing the label of real or vacuum based only on the threshold of the filter. The ambiguity in the link between imprint and imprinter leaves space to strategies that lead to the expansion of the content information of the image that can even influence the interpretation process.

If we associate a three-dimensional Cartesian space to the image (the usual physical space), each orthogonal axis becomes a source of dichotomies providing an amplification of the proper intellectual space (for instance proximity/distance, left/right, or in the axis orthogonal to the plane of the picture depth and height). Sometimes some dichotomies are introduced due to their anachronism: as is the case of wave/particle dichotomy or the association of a third dimension to images from a bidimensional universe. Adding imagination when interpreting something objective or introducing a third dimension to a flat world is a contradiction that can appropriately interweave in dialectics, overcoming the traditional deductive approach based on cause-and-effect relations.

Time is not a mere dichotomy, in fact its nature can be interpreted as a dualism in which the arrow and the cycle coexist, depending on the subjectivity of the observer.
Zeev Katvan

At this point of the dissertation about images the introduction of additional dimensions associated to cause-and-effect relations represented with arrows becomes spontaneous. The first arrow experienced is the thermodynamic arrow of time that progresses from order to chaos. More familiar is the biological arrow of time, which relates to the evolution of an organism with time (birth-death). In evolutionary processes, as time progresses the elements of the process irreversibly change. Life has evolved from simple (unicellular) to complex (multicellular); this growth is the path from chaos to order. A more specific arrow is the electromagnetic arrow of time. In the field of electromagnetism based on Maxwell's equations, only retarded waves are considered to be scientific since fundamental laws of physics can explain only retarded waves and not their counterpart. As a consequence we have a breakdown of time symmetry based on and originated by the principle of causality. This particular nature's phenomenon is logically meaningful by considering its one-sided behavior, i.e., irreversible advancement.

Linear time turned out to be insufficient for describing nature at its fundamental level. In the arrow of time every individual (thinking element) is different from the others. In cyclical time all the individuals of the population are to be considered equal (indiscernible) and interacting with each other.

A physical system where the dualistic nature of time emerges is the single-pulsed laser: in the presence of an external energy supply, two populations of atoms and photons interact, and while atoms can undergo a periodic activity of excitation and deexcitation (time's cycle), the whole laser system experiences an irreversible unidirectional evolution characterized by the emission of the photon pulse (time's arrow).

Looking at water flowing in a river or in a fountain (Fig. 20), it can happen to identify a main direction and several vortices, circular structures created by water moving in a direction opposite to the mainstream. This image on one hand emphasizes the importance of the observation scale (vortices are local phenomena and sometimes objects moving along the arrow of the waterway) and on the other hand can be useful to understand the complex relationship between linear and cyclic time.

Fluid turbulence is a multiscale phenomenon characterized by irregularity, strong vorticity, and rapid mixing. When the observer establishes a connection between an image/photograph and a natural process, often it is generated as a driving force that sometimes can be deceitful and lead to misinterpretation (Kemp 2006). In common life we face a lot of examples that induce us to think about natural processes. For example, when we look at an inanimate object as a manhole cover with robinprints, it is a static object that represents the passage of robins. Or again, when looking at animal tracks or to a photograph of animal tracks, the observer faces a static situation, but the human brain associates a temporal evolution to the image creating a story (a narration). Establishing a before and after implies assumptions regarding the existence of an animal, at least one, and the time of impression of the track. This brain function which operates trying to solve ambiguities can seem far from the interpretation of electron microscopy images. Actually it is not, as in the case of electron mirror images where electron trajectories are difficult to be described. The problem of trajectories is present also in images of animal tracks where identifying which tracks are part of the same pathway or portion of trajectory and which are part of other trails is complicated. This problem is difficult to be solved in the presence of a single impressor (animal) and gets intricated when the impressors are more than one.

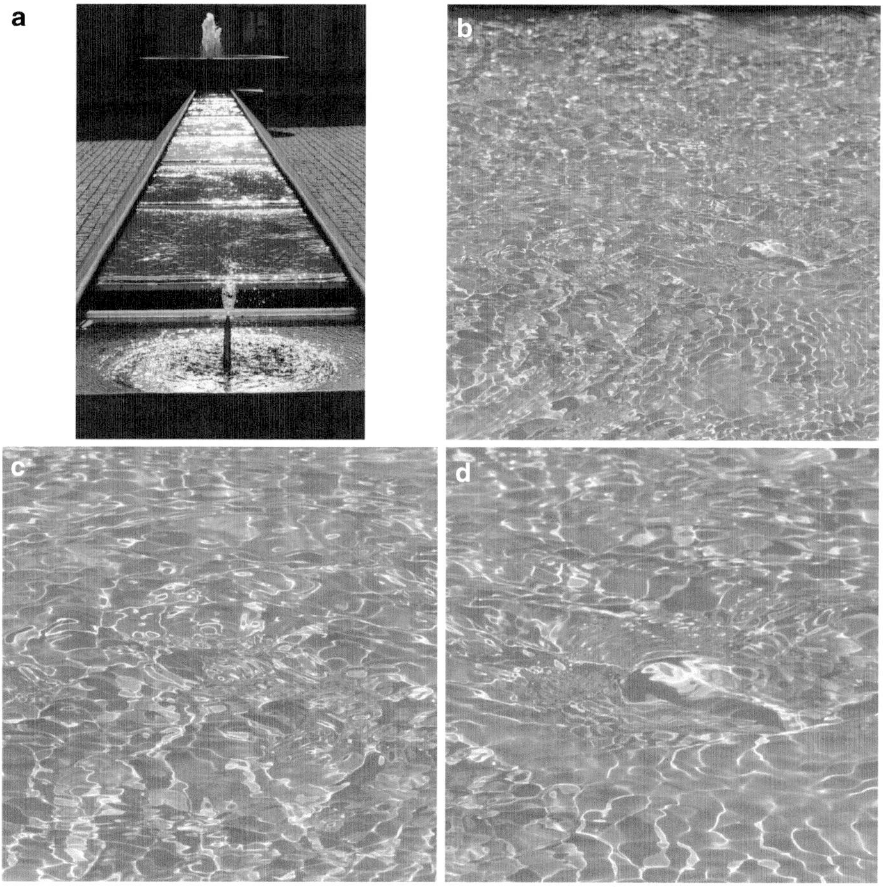

Fig. 20 Appearance of turbulence according to different observation scales. (**a**) Fountain in Erba, Como, Italy, by the architect Marco Castelletti. (**b**, **c**, **d**) Depending on the choice of the observer, different magnifications show chaotic changes in the turbulent flow

A me pare che il tempo – o il cambiamento suo sinonimo –
sia un mistero terrificante, forse il più grande di tutti i misteri.
It seems to me that time - or the change its synonym - is a terrifying mystery, perhaps the greatest of all mysteries.
Martin Gardner
(Authors' translation)

Feynman's sum-over-histories approach suggested to view time as a labyrinth of ever bifurcating possibilities. Each point in time has many branches towards the future and roots into the past by merging and splitting processes (Halpern 2018).

A labyrinth is to be understood as a bunch of paths that is characterized by the intentionality of the route and its systematic nature (Angelino 2003). As already mentioned, the labyrinth can be represented by an ensemble of broken lines that cross themselves, create nodes, diverge, and create loops. One of labyrinth's main features is the multi-dimensionality, a characteristic that introduces the possibility of chaotic pathways.

When considering a labyrinth and all its dynamics time becomes fundamental. Being a metaphor for knowledge, the labyrinth best represents nonlinear processes. Our investigation based on electron microscopy images of bacterial cells incubated with polyurethane has led to the instauration of an intricate neverending story. The interaction via infection dynamics between bacteria loaded with polyurethane nanoparticles and eukaryotic cells represents a complex web labyrinth with multi-faceted characters. The combination (cooperation/competition) of all the possible pathways and their temporal evolution can be represented in a map. This reticular scheme seems to have in common several analogies with the metropolitan urban model and therefore, as we will discuss later, with metropolitan connections (connections between geographic points of the city).

An example given by the investigation technique used to shed light on the Genchi case shows how the connection between spatial data (cellular cells), and calling and called telephone numbers contributed to building a structure of nodes and lines used to extracting determinant information. Due to legal laws, the content of the calls was excluded and the investigation was basded only on the geometric spatial-temporal graphical representations (Montolli 2009).

Perception derived from the encounter of word and image should render visible what is invisible using literary description of visual representation. The articulation of a visually narrative style is a balance between observation and description, pattern, and design and results in an archetype of the map, in which temporal and spatial totality are included. Our images are obviously static and time seems not to have a role. Making the same reasoning as per the images of animals' tracks, analyzing the spatial disposition and the morphology of membrane vesicles and nanoparticles, we can investigate all their possible routes. This process allow us to transform the initial spatial information in a sequence of temporal events, hence associating a time arrow to the single vesicles. This process is also reversible, so that the time arrow can be transformed into a cycle. This is the case of endocytosed vesicles (early endosome, late endosome) and of endocytosed/exocytosed nanoparticles (Huotari and Helenius 2011).

Considerations about the reading of images and the attribution of temporal orders that justify the image itself lead us to consider the inevitable contraposition between the reversibility *tout court* of classical mechanics and the one more sophisticated of quantum mechanics, and the irreversibility of the biological, physiological, and medical realities. Time and its properties are among the funding elements of relativity. Actually two points are usually neglected in common language (sometimes even in the scientific one): first of all what we handle are always time intervals and not single points of the temporal axis; secondly we skip any discussion about continuity or discreteness of the variable time itself—although the major part of solutions of problems are based on computational resources and therefore technically require a discretization process.

Although relativity attempts to unify time and space into one seamless object, time has a special nature that can manifest in the difference between time-translation symmetry and other symmetries. Spontaneous symmetry seems to suggest that almost any symmetry can be broken. However, recently it was shown that in periodically driven closed quantum systems, *discrete* time translation symmetry is not so different from other symmetries. A definition of discrete time-translation symmetry breaking allows distinctions between distinct nonequilibrium phenomena that superficially appear similar and exhibit oscillations with unexpected periods such as period-doubled nonlinear dynamical systems, mode-locked laser, parametric down-conversion, NMR (Nuclear Magnetic Resonance) spin echo, Belousov-Zhabotinsky reaction, convection cells, and AC (Alternating Current) Josephson effect (Yao and Nayak 2018). On the other hand, long Josephson junctions, which can be seen as a linear chain of coupled torsion pendula, are a model for time crystals. Once again we find ourselves dealing with complex systems that reveal their dynamics by means of the study of the behavior of a set of coupled pendula. The analogy between the dynamical equation describing a small Josephson junction and that describing a simple plane pendulum subjected to viscous damping and an applied mechanical torque is well known. The extension of this analogy to the long junction case consists of an elastic coupling between adjacent pendula; a very simple "pocket calculator" version can be easily constructed by inserting dressmaker's pins into a rubber band at regular intervals. In spite of the development of high-speed digital computers in recent years, the mechanical model of the long junction remains still today an extremely effective stimulus for physical intuition (Yao and Nayak 2018; Parmentier 1993; Barone and Paterno 1982).

This brings us again to the processes that generate notions associated to images and their interpretation. Notions can be sensitive, sensible, and intelligible or purely intelligible. They concur to the pursuit of knowledge via foreknowledge (sense perception or imagination), demonstrative reason, and an intuitive science of things. Following this scheme, we organized the processes of data mining from an image in three steps: observation of the image, interpretation related to the knowledge of the observed sample available at the time of the observation, and analysis of the intelligible information. In this way, starting from the labyrinthine structure of the neverending story, we finally managed to build the map of the whole system, with the aim to investigate and identify the possible paths and endpoints of the nanoparticles.

Since we are facing a complicated system that needs to show the complexity of the neverending story in its completeness, we decided to recur to a common representation as a metro map, of easily consultation when we decide to go from point A to point B in a big city with more than one underground line. One of the most famous representations of metro maps is the one of the Tube; since it is not based on the real geography of London, it must be considered as a diagram (Graham-Smith 2018). The representation in the form of a map has a geometric value; moreover it has a big advantage because it can lead to answers and results even without the need to perform complex calculations. The integration map-diagram gives more information and great help in understanding the processes connecting different stations. The map is like a linking web between points where, starting from the combinations of a limited number of basic elements and using the combinatory *ars*, it is possible to generate a world of infinite complexity. In this case we took as an example the map of the Moscow metro line (Fig. 21).

A voyage based on a metro map, even though apparently simple and linear, can hide complexities. Similar structures can have chaotic characteristics in which even a small uncertainty in the initial conditions can lead to an exponentially growing uncertainty in the predictions of its future. Chaos occurs in real systems and explains their irregular oscillations limiting the predictability of their future, even though often chaotic systems have mostly predictable time evolutions with a little bit of superimposed chaotic noise (Ruelle 1994).

The scheme of the neverending story (Fig. 1 in Chap. "*Staphylococcus aureus* Scouts the Nanoworld") has now the form of a metro map (Fig. 22a, b). This

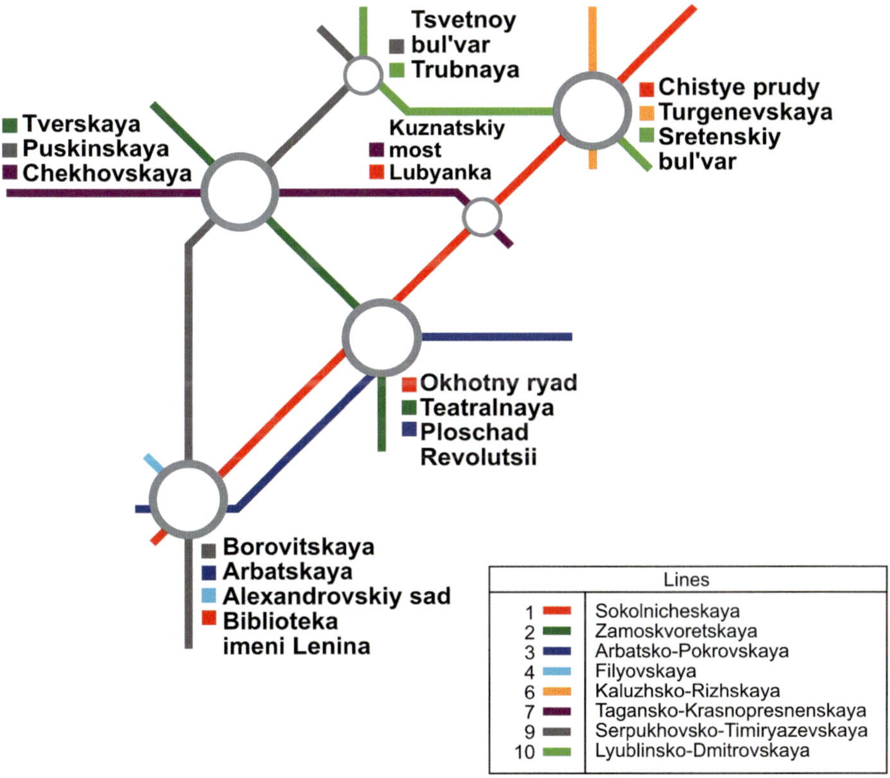

Fig. 21 Portion of the Moscow Metro map with nodes representing the crossroads of different lines. It is interesting to note that the same station, according to the different transit line, has a different name; it is as if a station can be named using several synonyms. This choice is not general, as for instance in Milan where one metro station, even if belonging to different lines, maintains the same name

complicated open system has a synergetic nature. The representation under the form of a map or a diagram evolves the representation from a disordered, or less ordered, status (Fig. 22a) to an ordered one (Fig. 22b).

In the labyrinthine structure shown in Fig. 22a the presence of several lines that cross themselves makes fundamental nonlinearities emerge. The possibility to have two different trajectories starting from a same point does not belong to classical mechanics and leads to instability and phase transitions. Some fundamental characteristics of the third type of labyrinth involve the presence of multiple loops and therefore the lengthening of the connection time intervals between the protagonists; moreover loops without output are possible and are characterized by infinite travel time independent of the distance and total uncertainty

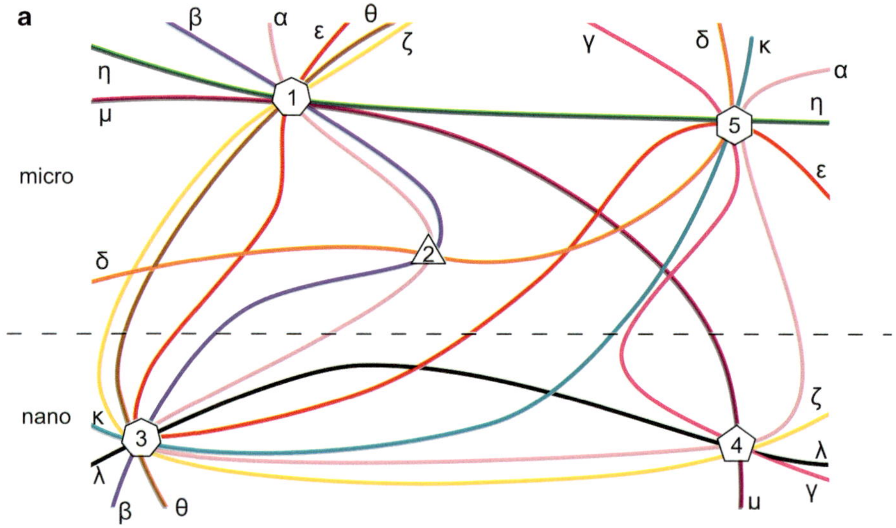

Fig. 22 (**a**) Representation of the *Staphylococcus aureus* neverending story through a map of curved lines that intersect. The image clearly illustrates the complex labyrinthine structure of the neverending story. Numbers represent nodes (stations), letters represent lines. Nodes: 1. *Staphylococcus aureus*; 2. bulk polyurethane; 3. polyurethane nanoparticles; 4. membrane vesicles; 5. eukaryotic host cells. *Lines*: α line—pink (1-2-3-4-5): neverending story backbone. β line—purple (1-2-3): adhesion, biodestruction, polyurethane nanoparticle formation. γ line—fuchsia (4-5): membrane vesicles and host eukaryotic cell interplay. δ line—orange (2-5): bulk material-host eukaryotic cell interaction. ε line—red (1-3-5): nanoparticles within *Staphylococcus aureus* in and out of host eukaryotic cells. ζ line—yellow (1-3-4): *Staphylococcus aureus*, polyurethane nanoparticle, and membrane vesicle interaction. η line—green (1-5): ordinary infectious process of *Staphylococcus aureus* on host eukaryotic cells. θ line—brown (1-3): *Staphylococcus aureus*-polyurethane nanoparticle interaction. κ line—light blue (3-5): polyurethane nanoparticle-host eukaryotic cell interaction. λ line—black (3-4): polyurethane nanoparticle-membrane vesicle interplay. μ line—plume (1-4): *Staphylococcus aureus*-membrane vesicle interplay. Beyond polyurethane nanoparticles, in light of the neverending story station number 3 can include also engineered or non-engineered nanoparticles deriving from processes not related with bacterial biodegradation, vesicles of different origin (from eukaryotic or bacterial cells), and viruses. (**b**) Representation of the *Staphylococcus aureus* neverending story (Fig. 22a) based on the technique in Fig. 21

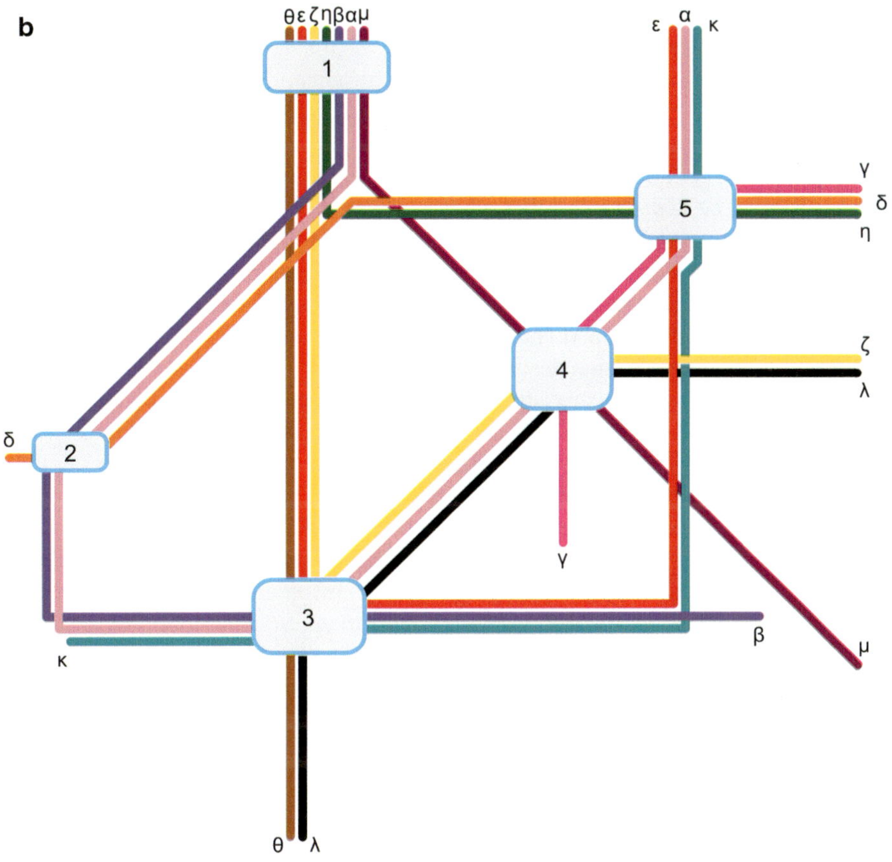

Fig. 22 (continued)

about the speed of connection. The complexity of the labyrinth and its specific properties help to highlight the important aspects of the neverending story that offer a multiplicity of approaches leading to further experiments. This labyrinthic representation can be a guide to pose the right questions to avoid ambiguities and confusing results.

> *We may acquire concepts by our perceptual experience of physical objects.*
> *But we would be mistaken if we thought that the concepts that we grasp were on the same level as the things we perceive.*
> *Marc Cohen*

All the elements discussed so far, integrating the aspects of the two cultures, help to improve the pathway of generation of the image and consequently its interpretation, preserving and integrating at the same time the roles of the microscopist and the researcher, taking away from the machine the responsibility of fallacious interpretations. The two following questions, posed through the language of the art historian Martin Kemp, enclose the contribution that our work aims to give to electron microscopy:

- What are the certainties we get from the figures?
- What do we really see? (Kemp 2006)

The two above questions are intertwined with an intricate series of considerations that in turn brings to other questions. When looking at images is it right to take into considerations details, single portrays of individual examples, sometimes also suppressing redundancies, instead of the model they refer to? In the image analysis it must be kept in mind that what an image portrays is the result of a particular dissection from a specific point of view, so it is mandatory for the image to be contextualized. In this way, through the analysis of images, processes could be better understood; according to this vision, images would act as points that need to be investigated, connected, and embedded in a context in order to acquire the concepts to build an overall visual description.

The production mechanisms and analysis of images remind us that time is a tyrant; we are slaves of time, our daily lives are characterized by the dualism between absence and presence, before and after, or in other words we can say that more is the time, stronger is the feeling that we miss it.

One way to ironically express this concept is the Parkinson's law that, as Dr. Didenko once stated, finds one special application in correspondence of deadlines for conferences or paper submission. Cyril N. Parkinson articulated the subjective perception of time in a comic (but not so much) manner in the first sentence of a humorous essay published in *The Economist* in 1955. The original sentence is formulated in a language common to thermodynamics "work expands so as to fill the time available for its completion" (Parkinson 1955) and is prone to be applied in numerous fields.

Parkinson's law has an intrinsic duplicity about time that recalls entropy (or order-disorder relations); apparently the law deals with just one time (better to say time lapse); in reality it implies the existence of two characteristic time lapses: on one hand the time useful to conclude a work that can be always contained in a minute (the shorter path) and on the other hand the available (assigned, often self-assigned) time, always the longest, at the end of the process. A better comprehension of Parkinson's law implications is supported by the language of electronics where the duty cycle (the ratio between the amount of activity time, when the system is working, and the sum of working and inactivity time) is introduced with reference to the total time necessary to complete a job.

The identity of the family name in Parkinson's disease and Parkinson's law occurs obviously by mere chance. Anyway this chance guides towards the existence of a phantastic liaison based on some kind of hidden mechanism mainly related to the concept of rigidity. Since the stimulus duration (typically the shortest time) points intuitively to the efficiency of the operator's brain, and the total time (typically the longest) refers to an inefficient (voluntary or involuntary) response to the stimulus, Parkinson's law looks like a simplified metaphor for the characteristic neural rigidity (a well-established characteristics of Parkinson's disease) of the operator.

Parkinson's law sprouted several corollaries, like: "If you wait until the last minute, it only takes a minute to do." This version perfectly fits for the authors of this book and their relation with the book itself: during the assigned time (self-assigned) and the "task" time, a lot of new ideas, thought, and themes of research emerged. Some ideas are still only partially developed and several of the original themes are still unfinished (not brought to the desired final state). Anyway, luckily in the end we were not facing a critical slowing down and what we now perceive is that the whole activity lasted only 1 minute. As we reached the end of this work the self-assigned deadline always postponed itself. This brings us to consider a particular type of time: the right time that identifies the critical moment to make the opportune action.

In our feelings this right time is suited for the memory of a dear person, Lyubov, a woman that perfectly represented the female figure in the Russian culture, and that had in herself a unique accord between a sense of tradition and a genuine freedom of thought, characteristics that made her a precious workmate. Inevitably the memory of Lyubov brings to our minds both the words of Pushkin "Where is the Moscow of hundred golden domes?" (Pushkin, 1815) and Kandinsky's "Red Square in Moscow" (1916) where he depicted his native city that he felt as a dear person. In the artist's view the personal traditional values and spiritual freedom together with the complexities and contradictions merged into the idea he had of his native city, Moscow, in short: "white-stoned," "gold-crowned."

References

Aaroe M, Monaco R, Rivers R, Koshelets V, Mygind J. Quench-induced trapping of magnetic flux in annular Josephson junctions. J Phys: Conf Ser. 2008:97. https://doi.org/10.1088/1742-6596/97/1/012279.

Abbott EA. Flatland: a Romance of Many Dimensions. Open Road Media Sci-Fi & Fantasy; 2014.

Acebrón JA, Bonilla LL, Vicente CJP, Ritort F, Spigler R. The Kuramoto model: A simple paradigm for synchronization phenomena. Rev Mod Phys. 2005;77:137. https://doi.org/10.1103/RevModPhys.77.137.

Agarwal GS, Shenoy SR. Observability of hysteresis in first-order equilibrium and nonequilibrium phase transitions. Phys Rev A. 1981;23:2719. https://doi.org/10.1103/PhysRevA.23.2719.

Alighieri D. The Divine Comedy: Inferno, Purgatorio, Paradiso. Penguin Classics; 2014.

Amodeo E, "Calvino e l'identità", 2013.

Andersen HC, "The shadow", 1847.

Anderson PW. Special effects in superconductivity. In: Caianello ER, editor. The Many-Body Problem, vol. 2. Elsevier; 2012. p. 113.

Angelino MI. I labirinti: il dramma del percorso. Doctor Virtualis. 2003:2. https://doi.org/10.13130/2035-7362/44.

Anon. "Why is Turing's halting problem unsolvable?", Scientific American n.d., retrievable at: https://www.scientificamerican.com/article/why-is-turings-halting-pr/#:~:text=A%20key%20step%20in%20showing,computer%20program%20will%20eventually%20halt.

Aquinas T. Summa Totius Theologiae. Anisson & Posuel; 1701.

Arbatsky DA, "The certainty principle II: proof of the uncertainty principle", arXiv: Quantum Physics, 2006.

Arecchi FT. High frequency. 1966;35(1):28–41.

Askar'yan, GA. "Cerenkov Radiation and Transition Radiation from Electromagnetic Waves". Journal of Experimental and Theoretical Physics. 1962;15(5):943–6.

Baker GL, Blackburn JA. The pendulum: a case study in physics. Oxford University Press; 2008.

Bandettini A. Dario Fo: "Perché il mio grammelot è una vera scienza". La Repubblica. 2014;

Baptist G. review of Nancy JL, "Le Regard du portrait", Galilée, 2000. http://www.kainos.it/Pages/recensione%20Nancy.html.
Barnes J. The complete works of Aristotle. Princeton University Press; 1991.
Barone A, Paterno G. Physics and applications of the Josephson effect. Wiley; 1982.
Barrow JD. The Book of Universes. Vintage; 2012.
Barthes R. "Mythologies", Les Lettres nouvelles 1957.
Barthes R. La chambre claire. Gallimard; 1980.
Bartocci E, Corradini F, Merelli E, Tesei L. Model checking biological oscillators. Electron Notes Theor Comput Sci. 2009;229(1):41–58. https://doi.org/10.1016/j.entcs.2009.02.004.
Becerra L, Barnes R, "Visual curiosities and mathematical paradoxes", 2010, retrievable at: https://plus.maths.org/content/visual-curiosities-and-mathematical-paradoxes. 2010.
Bednorz JG, Müller KA. Perovskite-type oxides—the new approach to high-T c superconductivity. Rev Mod Phys. 1988;60:585. https://doi.org/10.1103/RevModPhys.60.585.
Beian LA. Legal translations between linguistic principles and pragmatic context. Redefining Community in Intercultural Context. 2013;2(1):203–8.
Belhaj M, Jbara O, Odof S, Msellak K, Rau EI, Andrianov MV. An anomalous contrast in scanning electron microscopy of insulators: the pseudo-mirror effect. Scanning. 2000;22:352–6. https://doi.org/10.1002/sca.4950220603.
Bello M. Curator and Head of "Arts at CERN", n.d. retrievable at: https://flash---art.com/article/arts-at-cern-monica-bello/.
Belpoliti M. L'Affaire Moro: anatomia di un testo. In: Vecellio V, editor. L'uomo solo: l'affaire Moro di Leonardo Sciascia. La vita Felice; 2002.
Benvenuti G. «Un solo nome»: Manzoni in Sciascia. Studium. 2017;113(6):929–42.
Berra G. Le teste 'composte' e 'reversibili' dell'Arcimboldo e la nascita della natura morta. In: Ferino-Pagden S, editor. Arcimboldo. Skira; 2017.
Bertamini M, Spooner A, Hecht H. Naive optics: Predicting and perceiving reflections in mirrors. J Exp Psychol Hum Percept Perform. 2003;29(5):982–1002. https://doi.org/10.1037/0096-1523.29.5.982.
Bhaskar R. On the possibility of social scientific knowledge and the limits of naturalism. J Theory Soc Behav. 1978;8(1):1–28.
Black M. A Companion to Wittgenstein's Tractatus. Cambridge University Press; 1964.
Bokulich A, Parker W. Data models, representation and adequacy-for-purpose. Eur J Philos Sci. 2021;11:31. https://doi.org/10.1007/s13194-020-00345-2.
Bonifacio R, Casagrande F, Milani M. Superradiance and superfluorescence in Josephson junction arrays. Phys Lett A. 1984;101(8):427–31. https://doi.org/10.1016/0375-9601(84)90620-0.
Borges JL. The garden of forking paths. Penguin Classics; 2018.
Böhm David. "A Suggested Interpretation of the Quantum Theory in Terms of" Hidden" Variables. I." Physical Review. 1952;85(2):166–179.
Bragdon CF. A primer of higher space. The fourth dimension. Manas Press; 1913.
Brecht B. Life of Galileo. Methuen Publishing Ltd; 1960.
Breuker H, Drevermann H, Grab C, Rademakers AA, Stone H. Tracking and imaging elementary particles. Scientific American. 1991;265(2):58–63. https://www.jstor.org/stable/24938680
Brinkman BA. Hysteresis and phase transitions when grasping objects of different sizes. 2010 retrievable at https://guava.physics.uiuc.edu/~nigel/courses/563/Essays_2010/PDF/brinkman1.pdf
Brokate M, Sprekels J. Hysteresis and phase transitions. Springer Science & Business Media; 1996.
Bruno G. "Il Sigillo dei sigilli. I diagrammi ermetici. Giordano. Bruno filosofo e pittore", Mimesis 2015.
Buck J, Buck E. Mechanism of Rhythmic Synchronous Flashing of Fireflies: Fireflies of Southeast Asia may use anticipatory time-measuring in synchronizing their flashing. Science. 1968;159(3821):1319–27. https://doi.org/10.1126/science.159.3821.1319.
Caldirola P. Dalla microfisica alla macrofisica. Edizioni scientifiche e tecniche Mondadori. 1974;
Calvino I. La follia del mirino. Il contemporaneo; 1955.
Calvino I. Il viandante nella mappa. Collezione di sabbia. 1984a:21–8.

Calvino I. The adventure of a photographer. In: Calvino I, editor. Difficult loves. Harcourt Brace Jovanovich; 1984b.
Calvino I. Mr. Palomar. Martin Secker & Warburg Ltd; 1985.
Calvino I. Why read the classics? Penguin Classics; 2009.
Carbone M. In: Cassirer E, editor. Eidos ed eidolon: il problema del bello e dell'arte nei Dialoghi di Platone. Raffaello Cortina; 2009.
Carofiglio G. La versione di Fenoglio. Einaudi; 2019.
Cassirer E. Myth of the State. Yale University Press; 1946.
Cassirer E, editor. Eidos ed eidolon: il problema del bello e dell'arte nei Dialoghi di Platone. Raffaello Cortina; 2009.
Cassirer E. The Warburg Years, 1919-1933: essays on language, art, myth, and technology. Yale University Press; 2019.
Cederboum N. "Self-Portrait"–a study of the 'self': a quest for the creation and the development of the 'self' through a 'chain of observations'. Dissertation for the Anglia Ruskin University; 2009.
Cellucci C. Rethinking knowledge. Metaphilosophy. 2015;46(2):213–34. https://www.jstor.org/stable/26602301
von Chamisso A. The extraordinary story of Peter Schlemihl. 1814.
Chaturvedi S, Marmo G, Mukunda N, Simon R, Zampini A. The Schwinger representation of a group: concept and applications. Rev Math Phys. 2006;18(08):887–912. https://doi.org/10.1142/S0129055X06002802.
Chiappori A. Le immagini difficili. Considerazioni semiotiche sul ruolo della fotografia in "L'avventura di un fotografo" di Italo Calvino. In: Leone M, editor. Lexia. Aracne: Immagini efficaci; 2014. p. 17–8.
Clark TD. Electromagnetic properties of point-contact Josephson-junction arrays. Phys Rev B. 1973;8:137. https://doi.org/10.1103/PhysRevB.8.137.
Cooper B. The first black hole image: what can we really see? The Guardian. 2019;
Cooper LN. Bound electron pairs in a degenerate Fermi gas. Phys Rev. 1956;104:1189. https://doi.org/10.1103/PhysRev.104.1189.
Costato M, Degasperis A, Milani M (Eds.), "National workshop on nonlinear dynamics". Compositori, Bologna, Italy 1995.
Costato M, Milani M, Spinoglio L. Quantum mechanics: a breakthrough into biological system dynamics. Bioelectrochem Bioenergetics. 1996;41(1):27–30. https://doi.org/10.1016/0302-4598(96)01922-8.
Courtney SJ. The salt-point: Kairos emergent from chaos. J Jungian Sch Stud. 2019;14:1–12. https://doi.org/10.29173/jjs3s.
Cowan N. What are the differences between long-term, short-term, and working memory? Prog Brain Res. 2008;169:323–38. https://doi.org/10.1016/S0079-6123(07)00020-9.
Crawford K. Atlas of AI. Yale University Press; 2021.
Cubitt S, "Suture", 2014.
Cusano N. I dialoghi dell'idiota. Olschki; 2003.
Davydov AS. Quantum mechanics. Pergamon Press; 1976.
Del Giudice E, Smith C, Vitiello G. Magnetic flux quantization and Josephson Systems. Phys Scr. 1989;40:786–91. http://iopscience.iop.org/1402-4896/40/6/017
Di Marco R. "Materialismo e funzione espressiva", Materialismo e pratica artistica. Scritti della rivista Cinetique (71-74), L'unità del sapere n.8, Lavoro Liberato 1976.
Di Rienzo A, Rogovin D, Scully M, Bonifacio R, Lugiato L, Milani M. Superconductivity and quantum optics. In: Arecchi FT, Bonifacio R, Scully MO, editors. Coherence in spectroscopy and modern physics, NATO advanced study institutes series. Boston, MA: Springer; 1978. p. 37. https://doi.org/10.1007/978-1-4613-2871-1_12.
Ditzinger T, Haken H. Oscillations in the perception of ambiguous patterns a model based on synergetics. Biological Cybernetics. 1989;61:279–87. https://doi.org/10.1007/BF00203175.
Dobroljubov N. Che cos'è l'oblomovismo. Il contemporaneo; 1859.
Dolling LM, Gianelli AF, Statile GN, editors. The tests of time. Readings in the development of physical theory. Princeton University Press; 2003.

Dostoevskij F. Il sosia. Rizzoli. 2009;
Durand F, "The art and science of depiction", n.d. retrievable at: https://people.csail.mit.edu/fredo/gbRealisme.pdf.
Dutta S, Cooper NR. Critical response of a quantum van der Pol oscillator. Phys Rev Lett. 2019;123(25):250401. https://doi.org/10.1103/PhysRevLett.123.250401.
Dyson F. The scientist as rebel. Am Math Month. 1996;103(9):800–5. https://doi.org/10.1080/00029890.1996.12004822.
Eastham PR, Szymanska MH, Littlewood PB. Phase-locking in quantum and classical oscillators: polariton condensates, lasers, and arrays of Josephson junctions. Solid State Commun. 2003;127(2):117–22. https://doi.org/10.1016/S0038-1098(03)00338-7.
Eco U. Semiotica e filosofia del linguaggio. Einaudi; 1984.
Eco U. Foucault's pendulum. Mariner Books; 2007.
Eco U. Sugli specchi e altri saggi. Il segno, la rappresentazione, l'illusione, l'immagine. Bompiani. 2015;
Eco U. Trattato di Semiotica Generale. La Nave di Teseo; 2016.
Eigen M, Schuster P. The hypercycle. Naturwissenschaften. 1978;65:7–41. https://doi.org/10.1007/BF00420631.
Einhäuser W, Martin KA, König P. Are switches in perception of the Necker cube related to eye position? Eur J Neurosci. 2004;20(10):2811–8. https://doi.org/10.1111/j.1460-9568.2004.03722.x.
Eliade M. The myth of the eternal return: or, cosmos and history. Bollingen Foundation; 1992.
Erega A, Curia R, Milani M. Interactions between nanoparticles and bacterial structures as seen through software enhanced analysis of electron microscopy images. In: Méndez-Vilas A, editor. Microscopy and imaging science: practical approaches to applied research and education. Formatex Research Center Publishing; 2017. p. 587–94.
Escher MC. The graphic work of M.C. Escher. Macmillan; 1972.
Feynman RP. QED: the strange theory of light and matter. Princeton University Press; 2006.
Firestein S. Ignorance: how it drives science. Oxford University Press; 2012.
Flusser V. Towards a philosophy of photography. Reaktion Books; 2000.
Fo D. Manuale minimo dell'attore. Einaudi; 1997.
Fröhlich H. Long-range coherence in biological systems. La Rivista del Nuovo Cimento (1971-1977). 1977;7(3):399–418.
Fröhlich H, Kremer F, editors. Coherent excitations in biological systems. Berlin Heidelberg: Springer-Verlag; 1983.
Gadamer HG. Che cos'è la verità? I compiti di un'ermeneutica filosofica. Rubbettino Editore. 2012;
Galatino N, "Verità. Solo i coraggiosi la ricercano", Il Sole 24 Ore, 07/15/ 2018.
Garmire E. Chiao RY, Townes CH. "Dynamics and Characteristics of the Self-Trapping of Intense Light Beams". Physical Review Letters. 1966;16(9):347
Gautier T "Avatar", La Haye, Les Héritiers Doorman 1857.
Glauber RJ. Classical behavior of systems of quantum oscillators. Phys Lett. 1966;21(6):650–2. https://doi.org/10.1016/0031-9163(66)90111-9.
von Goethe JW, "Faust", 1808.
von Goethe JW. Collected works of Johann Wolfgang von Goethe. Delphi Classics; 2013.
Gombrich EH. Shadows: the depiction of cast shadows in Western Art. Yale University Press; 1995.
Gombrich EH, Hochberg J, Black M. Art, perception, and reality. JHU Press; 1972.
Gončarov I. Oblomov. Rizzoli; 2006.
Gould SJ. Time's arrow, time's cycle: myth and metaphor in the discovery of geological time. Harvard University Press; 1987.
Graham-Smith D. "The history of the tube map", 2018 retrievable at: https://londonist.com/2016/05/the-history-of-the-tube-map.
Gregory RL. Perceptions of knowledge. Nature. 2001;410:21. https://doi.org/10.1038/35065174.
Haken H. Laser Theory. In: Genzel L, editor. Light and Matter Ic. Berlin, Heidelberg: Springer; 1970. p. 5/25/2/2c. https://doi.org/10.1007/978-3-662-22091-7_1.

Haken H, editor. Synergetics. Cooperative phenomena in multi-component systems. Vieweg+Teubner Verlag Wiesbaden; 1973. https://doi.org/10.1007/978-3-663-01511-6.
Haken H. Synergetics. An introduction. Nonequilibrium phase transitions and self-organization in physics, chemistry and biology. Berlin Heidelberg: Springer-Verlag; 1978. https://doi.org/10.1007/978-3-642-96469-5.
Halpern P. The Quantum Labyrinth. How Richard Feynman and John Wheeler revolutionized time and reality. Basic Books; 2018.
Haverkamp A. The memory of pictures: Roland Barthes and Augustine on photography. Comp Lit. 1993;45(3):258–79. https://doi.org/10.2307/1771504.
Hecht H, Bertamini M, Gamer M. Naive optics: acting on mirror reflections. J Exp Psychol Hum Percept Perform. 2005;31(5):1023–38. https://doi.org/10.1037/0096-1523.31.5.1023.
Henry M, "Voir l'invisible. Sur Kandinsky", PUF, 2014.
von Helmholtz H. Handbuch der physiologischen Optik. Hamburg und Leipzig: Leopold Voss; 1896.
Hizanidis J, Lazarides N, Tsironis GP. Chimera states in networks of locally and non-locally coupled SQUIDs. Front Appl Math Stat. 2019;5:33. https://doi.org/10.3389/fams.2019.00033.
Holmes OW, "The autocrat of the breakfast-table", 1858, retrievable at: https://www.ibiblio.org/eldritch/owh/abt03.html.
Holtzmann Kevles B. The interplay of art and science. Scientific American; 2007.
Hoskin K. The viewing self and the world we view: Beyond the perspectival illusion. Organization. 1995;2(1):141–62. https://doi.org/10.1177/135050849521007.
Hubel DH, Wiesel TN. Brain mechanisms of vision. Sci Am. 1979;241(3):150–63. https://www.jstor.org/stable/24965293
Huizinga J. La crisi della civiltà. Einaudi; 1935.
Huotari J, Helenius A. Endosome maturation. The EMBO journal. 2011;30(17):3481–500. https://doi.org/10.1038/emboj.2011.286.
Hyland P. Reading as rewriting: Miguel de Unamuno, Jorge Luis Borges and the Quijote. Doctoral dissertation, University of Cambridge 2019.
Janson AF. "The convex mirror as Vanitas symbol", Source. Notes Hist Art. 1985;4(2/3):51–4. https://doi.org/10.1086/sou.4.2_3.23202426.
Jones D. Seeing reason: how to change minds in a 'post-fact' world. New Scientist. 2016;
Josephson BD. Possible new effects in superconductive tunnelling. Phys Lett. 1962;1(7):251–3. https://doi.org/10.1016/0031-9163(62)91369-0.
Kadanoff LP. Roads to chaos. In: Kadanoff LP, editor. From order to chaos: essays: critical, chaotic and otherwise; 1993. p. 391–8. https://doi.org/10.1142/9789812798763_0025.
Kandinsky VV. Point and line to plane. New York: The Solomon R Guggenheim Foundation; 1947.
Kandinsky VV. Lo spirituale nell'arte. SE; 2005.
Karpatschof B. Human activity. Contributions to the anthropological sciences from a perspective of activity theory. Dansk Psykologisk Forlag 2000.
Katvan Z. Reflection theory and the identity of thinking and being. Stud Sov Thought. 1978;18(2):87–109. https://www.jstor.org/stable/20098785
Kemp M. From Mimesis to Fantasia: the Quattrocento vocabulary of creation, inspiration and genius in the visual arts. Viator. 1977;8:347–98. https://doi.org/10.1484/J.VIATOR.2.301573.
Kemp M. The science of art: optical themes in Western Art from Brunelleschi to Seurat. Yale University Press; 1992.
Kemp M. Journey into space. Nature. 1999;400:823. https://doi.org/10.1038/23602.
Kemp M. Immagine e verità. Net. 2006;
Knorr WR. On the principle of linear perspective in Euclid's optics. Centaurus. 1991;34(3):193–210. https://doi.org/10.1111/j.1600-0498.1991.tb00694.x.
Koffka K. Principi di psicologia della forma. Bollati Boringhieri; 2006.
Kuramoto Y, editor. Chemical oscillations, waves, and turbulence. Dover Press; 2003.
Leary WE. Peary made it to the pole after all, study concludes. The New York Times. 1989;
Lecourt D. Une crise et son enjeu: essai sur la position de Lénine en philosophie. Maspero. 1973;
Lenin VI. Materialismo ed empiriocriticismo. PGreco. 2021;

Leonelli S. Philosophy of biology: the challenges of big data biology. Elife. 2019;8:e47381. https://doi.org/10.7554/eLife.47381.
Lévy-Leblond J-M. La velocità dell'ombra. Ai limiti della scienza. Codice; 2007.
Likharev KK. Superconducting weak links. Rev Mod Phys. 1979;51:101. https://doi.org/10.1103/RevModPhys.51.101.
London J, "The shadow and the flash", 1903.
Loos A, "Symbolic, real, imaginary", 2002, retrievable at: https://csmt.uchicago.edu/glossary2004/symbolicrealimaginary.htm.
Luk'yanov AE, Spivak GV, Gvozdover RS. Mirror electron microscopy. Sov Phys Uspekhi. 1974;16(4):529–52.
Lumer L, Zeki S. La bella e la bestia: arte e neuroscienze. Laterza, 2011.
Luque A. Borges in Sicily. The Armchair Traveller at the Bookhaus; 2017.
Lynch I, Salvati A, Dawson KA. What does the cell see? Nature nanotechnology. 2009;4:546–7. https://doi.org/10.1038/nnano.2009.248.
Lyubchenko I. The Proto-Pixel Art of Malevich and Kandinsky: black square, its digital descendant and neo-vitalist impulse. 2014.
Mace JH. Involuntary autobiographical memories are highly dependent on abstract cuing: The Proustian view is incorrect. Appl Cogn Psychol. 2004;18(7):893–9. https://doi.org/10.1002/acp.1020.
Macherey P, Balibar E. Literature as an ideological form: some Marxist propositions. Oxford Literary Review. 1978;3(1):4–12. http://www.jstor.org/stable/44244885
Malvaldi M, Leporini D. Capra e calcoli. L'eterna lotta tra gli algoritmi e il caos. Laterza; 2014.
Mansfield E, Kaiser DL, Fujita D, Van de Voorde M, editors. Metrology and standardization for nanotechnology: protocols and industrial innovations. John Wiley & Sons; 2017.
Manukyan K. The poet and his readers: three lyrics and an unfinished story of Alexander Pushkin. Pushkin Rev. 2009;12:17–43. https://www.jstor.org/stable/43783204
Marramao G. Kairós. Apologia del tempo debito. Laterza; 1992.
Mason J, Kaszor P, Bourassa CM. Perceptual structure of the Necker cube. Nature. 1973;244:54–6. https://doi.org/10.1038/244054a0.
Mathieu V, Rossi P. Scientific culture in the contemporary world. Scientia. 1979;
de Maupassant, "Lui?", 1883.
McCurry S. The iconic photographs. Phaidon. 2012;
Melchior-Bonnet S. Storia dello specchio. Dedalo. 2002;
Milani M, Abdul-Wahab HN, Abbood TH, Savoia C, Tatti F. "Rear window": looking at charged particles hitting a charged target in a FIB/SEM. In: Méndez-Vilas A, Díaz J, editors. Microscopy: science, technology, applications and education. Formatex; 2010.
Milani M, Bonifacio R, Scully M. A model for bistability in a coherently driven Josephson junction. Nuovo Cimento Lettere. 1979;26:353–60.
Milani M, Del Giudice E, Doglia S, Vitiello G, Smith CW. Superconductive and Josephson-like behaviour of cells. La Radiologia Medica. 1991;81:51–5.
Mody CC. Between research and development: IBM and Josephson computing. Phys Today. 2016;69(10):32.
Montolli E. Il caso Genchi. Storia di un uomo in balìa dello Stato. Aliberti. 2009;
Motter AE, Campbell DK. Chaos at fifty. In: Pitici M, editor. The best writing on mathematics. Princeton University Press; 2014.
Nabokov VV. The Eye. Penguin Classics; 1992.
Nanay B. Mental imagery. Stanford Encyclopedia of Philosophy, 1997. Retrievable at: https://plato.stanford.edu/entries/mental-imagery/.
Necker LA. Observations on some remarkable phenomenon which occurs in viewing a figure of a crystal or geometrical solid. Lond Edinb Philos Mag J Sci. 1832;3:329–37.
Nicoli E. Italo Calvino e Lucrezio: dalla fase combinatoria alle Lezioni americane. In: Citti F, Pellacani D, editors. Ragione e furore. Pendragon: Lucrezio nell'Italia contemporanea; 2020.
Nicolis G, Prigogine I. Self-organization in non-equilibrium systems. From dissipative structures to order through fluctuations. Wiley; 1977.

Nielsen K, "Image", 2003. Retrievable at: https://lucian.uchicago.edu/blogs/mediatheory/keywords/image/.
Nilsen DL. Doppelgangers and doubles in literature: A study in tragicomic incongruity. Humor. 1998;11(2):111–33. https://doi.org/10.1515/humr.1998.11.2.111.
Nilsson AN. Brevity in Pushkin's Prose. Can Slavon Pap. 1987;29(2-3):152–64. https://doi.org/10.1080/00085006.1987.11091854.
Ogden B. Benjamin, Wittgenstein, and philosophical anthropology: a reevaluation of the mimetic faculty. Grey Room. 2010;19:57–73. https://www.jstor.org/stable/27809426
Oliveri M. La letteratura italiana nella pagine della critica. Paravia; 1959.
Panofsky E. Galileo as a critic of the arts: Aesthetic attitude and scientific thought. Isis. 1956;47(1):3–15. https://doi.org/10.1086/348450.
Parkinson CN. Parkinson's law. Economist. 1955;
Parkinson J. An essay on the shaking palsy. J Neuropsychiatry Clin Neurosci. 2002;14(2):223–36.
Parlagreco S. Il mistero del corvo. Il caso Di Pisa. SugarCo; 1990.
Parmentier RD. Solitons and Long Josephson Junctions. In: Weinstock H, Ralston RW, editors. The new superconducting electronics. Dordrecht: Springer; 1993. https://doi.org/10.1007/978-94-011-1918-4_7.
Pauli W. Psiche e Natura. Adelphi; 2006.
Perrault C. Cenerentola. Donzelli; 2011.
Peters BJ. Telescope+ mirror = reflections on the cosmos: Umberto Eco and the image of religion. Zygon. 2017;52(2):343–60.
Piana G, "Il canto del merlo", 2007.
Pino A, "Kandinsky e la musica", n.d. retrievable at: http://www.musil.it/eventi/Kandinsky/K--musica.htm.
Pirandello L. One, no one and one hundred thousand. Marsilio; 1992.
Pirandello L, "Right You Are! If You Think So", Createspace Independent Pub, 2015. Translated by Arthur Livingston.
Pirandello L. Right You Are (If You Think So): 'It is so. When YOU think so'. Stage Door. 2020;
Pistoletto M. Ominiteismo e demopraxia. Chiarelettere, 2017.
Plato, "Cratylus", ~ 360 BC n.d.-a
Previdi F, Milani M. Cellular-automaton approach to two-dimensional VCSEL array collective behaviour. Nuovo Cimento-Societa Italiana Di Fisica Sezione D. 1998;20:1625–34.
Prigogine I. Time, irreversibility and structure. In: Mehra J, editor. The physicist's conception of nature. Dordrecht: Springer; 1973. https://doi.org/10.1007/978-94-010-2602-4_29.
Prigogine I. From being to becoming: time and complexity in the physical sciences. W H Freeman & Co; 1981.
Proust M. Remembrance of things past. Faber & Faber; 2000.
Proust M. In search of lost time. Modern Library; 2003.
Pushkin AS, The Reminiscences in Tsarskoe Selo, 1815.
Raepple EM. "Experience Does Not Err" (Leonardo Da Vinci)-artwork as a mirror of nature. Philosophy Scholarship. 2009:7. https://dc.cod.edu/philosophypub/7
Rogers C. WHF Talbot, *The Pencil of Nature* and the camera. Hist Photogr. 2002;26(2):141–50. https://doi.org/10.1080/03087298.2002.10443288.
Röhm A, Lüdge K, Schneider I. Bistability in two simple symmetrically coupled oscillators with symmetry-broken amplitude-and phase-locking. Chaos: An Interdisciplinary Journal of Nonlinear Science. 2018;28(6):063114. https://doi.org/10.1063/1.5018262.
Ross T. Metaphor and paradox. Ga. L. Rev. 1989;23:1053.
Ruelle D. Where can one hope to profitably apply the ideas of chaos? Phys Today. 1994;47(7):24–30.
Rushmore S. Medical education and culture. Science. 1924;59(1539):569–71. https://doi.org/10.1126/science.59.1539.569.
Sacks O. The mind's eye. Knopf; 2010.
de Saint-Exupéry A. The little prince. Harcourt: Brace & World; 1943.
Scales JA, Snieder R. What is a wave? Nature. 1999;401:739–40. https://doi.org/10.1038/44453.
Scarpa D, Italo Calvino, Bruno Mondadori, 2005.

Schmahl FW, von Weizsäcker CF. Medicine and modern physics. The Lancet. 1998;351(9111):1291–2. https://doi.org/10.1016/S0140-6736(05)79360-0.
Schneider M. Calvino at a Crossroads: Il castello dei destini incrociati. PMLA. 1980;95(1):73–90. https://doi.org/10.2307/461734.
Schopenhauer A. On vision and color. Princeton Architectural Press; 2010.
Schurz G. Philosophy of science: a unified approach. Routledge; 2013.
Sciascia L. The Moro Affair. Granta Books; 2014.
Scott A. Stairway to the mind: the controversial new science of consciousness. Springer Science & Business Media; 2012.
Sini C. I segni dell'anima. Saggio sull'immagine. Laterza; 1989.
Snow CP. The two cultures and the scientific revolution. Cambridge University Press; 1959.
Snowden E. Permanent record. Pan Macmillan; 2019.
Solymar L. Superconducting tunneling and applications. London: Chapman and Hall; 1972.
Spirkin A. "Dialectical materialism", Themes in Soviet Marxist Philosophy, Springer, Dordrecht, 1975.
Sprawls P. Magnetic resonance imaging: principles, methods, and techniques. Medical Physics Publishing Corporation; 2000.
Stevenson RL. The strange case of Dr. Jekyll and Mr. Hyde. Harper UK 2010.
Strogatz S. Sync: the emerging science of spontaneous order. Penguin Press Science; 2004.
Tarasov LV. Basic concepts of quantum mechanics. Mir Publishers Moscow; 1980.
Tolstoj L. War and peace. StreetLib. 2016;
Tyndall J. In: von Goethe JW, editor. Collected works of Johann Wolfgang von Goethe. Delphi Classics; 2013.
Van der Pol B, Van Der Mark J. Frequency demultiplication. Nature. 1927;120:363–4. https://doi.org/10.1038/120363a0.
de Vidovich S. F. M. Dostoevskij. Il sosia. Rivista di Psicologia Analitica. 1984;30:46–63.
Villard R, Levay Z. Creating Hubble's technicolor universe. Sky Telescope. 2002;104(3):28.
Von Fieandt K. Über Sehen von Tiefengebilden bei wechselnder Beleuchtungsrichtung, 1938.
Von Neumann J. The computer & the brain. Yale University Press; 2012.
Wessen K. Not just a matter of time: the halting problem. 2016. retrievable at: https://plus.maths.org/content/not-just-matter-time-part-ii.
Wiggers KL, Britto AS, Heutte L, Koerich AL, Oliveira LES. Document image retrieval using deep features 2018 International Joint Conference on Neural Networks (IJCNN), 2018. https://doi.org/10.1109/IJCNN.2018.8489722.
Wilde O. The picture of Dorian Gray. Penguin Classics; 2019.
Winfree AT. Biological rhythms and the behavior of populations of coupled oscillators. J Theor Biol. 1967;16(1):15–42. https://doi.org/10.1016/0022-5193(67)90051-3.
Yao NY, Nayak C. Time crystals in periodically driven systems. Physics Today 71, 2018;40–47 https://doi.org/10.1063/PT.3.4020
Zeki S. The neurology of ambiguity. Conscious Cogn. 2004;13:173–96. https://doi.org/10.1016/j.concog.2003.10.003.
Zeki S. Splendori e miserie del cervello. L'amore, la creatività e la ricerca della felicità. Codice. 2010;
Zimmer H. Parametric amplification of microwaves in superconducting Josephson tunnel junctions. Appl Phys Lett. 1967;10(7):193–5. https://doi.org/10.1063/1.1754906.

Post scriptum

Not to interpret is impossible, as refraining from thinking is impossible.
Italo Calvino

The tribute to Dr. Didenko does not conclude with the end of our voyage in the world of bacteria and nanoparticle. Both Dr. Didenko's memory and teaching have enhanced our way of thinking and increased our sensibility, imagination, and *phantasia*; all this helped us to create connections between elements of the two cultures and embed them in the contemporary scientific context, as a continuous stimulus to our curiosity and will of research.

The preparation of this book gave rise to some ideas that open to new perspectives of the neverending story told so far.

Electron microscopy images are a complex structured ensemble of black and white points that can be thought of as a combination of presence and absence of tracks; the buildup of scientific knowledge would be achieved through the improvement of the information extraction from this kind of images. As we have discussed through the course of this book, there can be several issues associated to the interpretation of electron microscopy images that could potentially compromise their scientific value. In our opinion the need to remove as many ambiguities as possible from the reading of the images is necessary to be open to a less biased interpretation.

We will not be as cynical to state that all images recorded by a TEM are artifacts.
Elisabeth Mansfield

Taking the hint from several activities such science, literature, and philosophy where it is a quite common practice, we will recur to maps, diagrams, and schemes to help solve the ambiguity problem. In our particular case we will refer to the map of the city of Moscow and the CXEMA (scheme) of its subway (Fig. 1). These two different illustrations can be listed in the synonym set under the voices "map," "painting," and "diagram," each with an obvious and evident diverse

M. Milani et al., *Bacterial Degradation of Organic and Inorganic Materials*,
https://doi.org/10.1007/978-3-031-26949-3_7

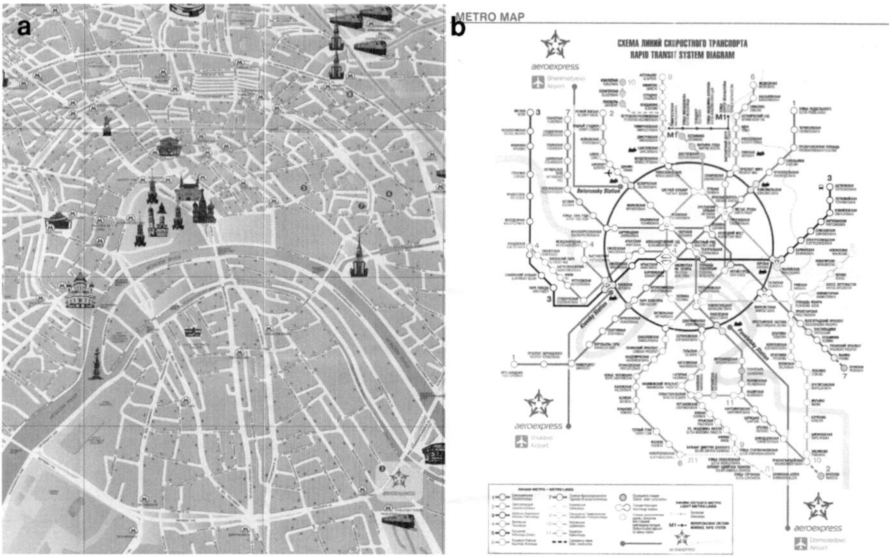

Fig. 1 Map of the city of Moscow (**a**) and scheme of its subway (**b**) from the free map "Moscow in your pocket" distributed in 2012

content of information and an intrinsic richness triggered by the situation the observer is in: spectator, reader, user, traveler, or explorer, a set of attitudes that merge altogether and are well described by Calvino in the character of Mr. Palomar (Calvino 1985).

Mr. Palomar's pursuit of knowledge takes place at different levels, and the interpretation becomes a fundamental operation constantly performed to give meaning to a reality otherwise devoid of its own significance. The interpretation of reality with one's own goggle lenses, ideas, and emotions is required to properly perceive reality itself (Bresciani Califano 2011). The critical difference between the fictional Mr. Palomar and a person living in the modern era is more or less the one that distinguishes *Homo sapiens* from *Homo videns*, where the first was used to observe the environment and subsequently deduce which actions to undertake, whereas the second is a typical character of today's society with poor abstraction and analytical skills, whose perception is reduced to simple seeing (Szymura 2009; Salamone 1999). *Homo videns* is a definition introduced by Giovanni Sartori, an Italian political scientist, to delineate the change that occurred after the introduction of television, a transition that did not mark an evolution but rather an involution for the human species (Sartori 2000). For the first time in history, images (visual messages) started to predominate and prevail over words (acoustic messages), destroying communication among human beings, mining the abstract thought that inevitably led to an intellectual regression and the consequent inability to distinguish virtual from real or true from false (Wright 2010; Bauman 2007; Hardin 2002; Ricci 2001).

Mr. Palomar, applying a naturalistic approach to knowledge as a response to the environment, tries to understand the phenomenological world through its observation, at the same time posing questions about the reliability of his perceptions (Cellucci 2015). Soon Palomar realizes that the truth is no longer the mere reflection of reality and that he needs to make choices about what to observe and what to exclude, convincing himself that reality is shaped by consciousness (Bresciani Califano 2011; Cannon 1985). The same choices need to be made when looking at an image that can lead to more than one interpretation; hence Mr. Palomar is the literature representation of the perfect approach a scientist that faces the analysis and interpretation of electron microscopy images should have.

The effects of the multiplicity of the available choices were already shown in Fig. 2 of chapter "*Staphylococcus aureus* Scouts the Nanoworld: A Neverending story": in the three panels the points derive from the same data sheet of coordinates in a Cartesian plane, whereas the connection lines, from which the three different structures emerge, are built by the observer. Once again it is evident that an image can contain more information and answers than the questions initially posed; this needs to be taken into account when investigating any kind of images and also electron microscopy ones. In our specific case it is related to an issue that occurred during the investigation of the distribution and trafficking of vesicles and nanoparticles inside a *Staphylococcus aureus* cell (Erega et al. 2017). Each exploration of some unknown sample, which can be done by the researcher or using dedicated software (both ways accompanied by issues), is always affected by uncertainty, indeterminacy, and arbitrariness in the fundamental operations of identification of structures, substructures, patterns, and connections.

In the case of a multicentered interpretation of a set of points where the observer can do multiple choices for the centers, the operation of reading an image leads to different selections and subsequently to the identification of different patterns—a process similar to the identification of constellations when looking to a starred sky or to the correlation of paths in endosome/lysosome dynamics in cells, from early to late endosome and vice versa (Huotari and Helenius 2011). Different examples of how it is possible to establish connections in totally different fields are represented in Fig. 2 that illustrates how apparently meaningless sets of points suggest potential paths and structures.

> *As we gradually tear the point out of its restricted sphere of customary influence, … the dead point becomes the living thing.*
> *Vasily V. Kandinsky*

Also in our investigation of polyurethane nanoparticles internalized by *Staphylococcus aureus*, the set of points highlighted different ways to establish connections. The spatial distribution of the membrane vesicles inside the bacteria (Fig. 2b) suggests the existence of correlations among the individual units sustained by organizing agents or processes. These correlations, appearing as structures or patterns, arise from elements and mechanisms at work in the cell activity such as the cytoskeleton, electromagnetic fields, chemical reactions, cytoskeletal dynamics, and transport phenomena.

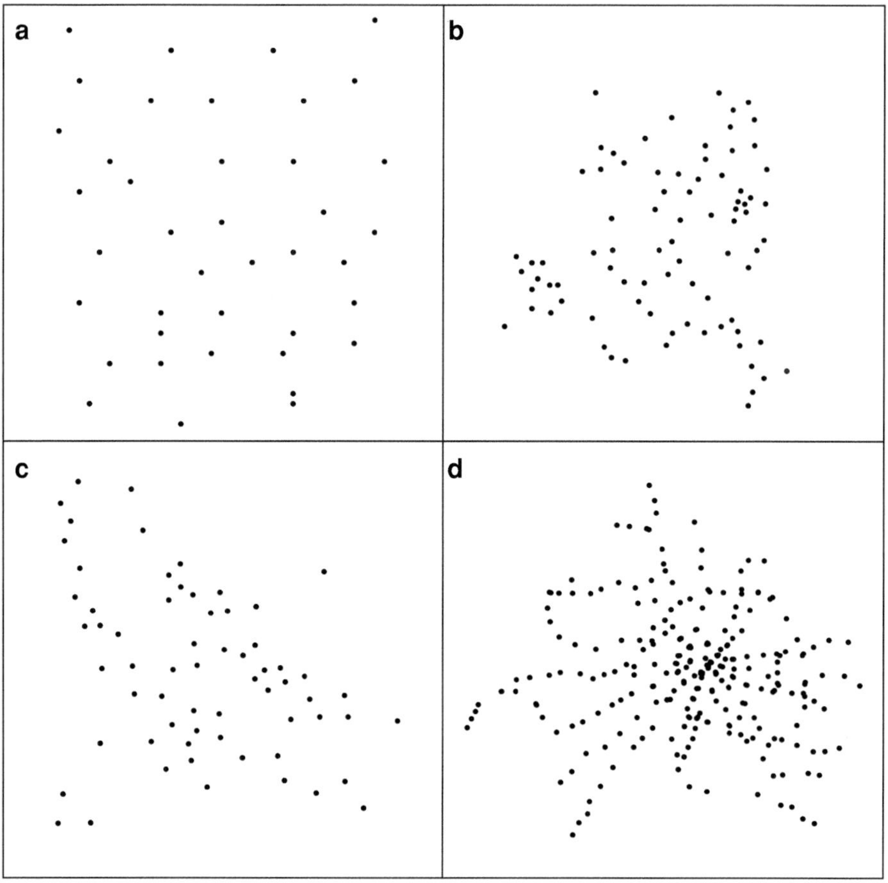

Fig. 2 Point ensemble's potential: (**a**) set of points associated to Fig. 2 of chapter "*Staphylococcus aureus* Scouts the Nanoworld: A Neverending Story"; (**b**) barycenters of vesicle distribution in *Staphylococcus aureus* TEM image, without any reference to their size or nanoparticle content, as seen in Fig. 1c in Erega et al. (2017); (**c**) authors' elaboration of Kandinsky's "Diagram 2—Point—Dissolution in progress (suggested diagonal **d**–**a**)," from *Point and Line to Plane* (Kandinsky 1947); (**d**) the spatial distribution of the 264 Moscow metro stations (https://en.wikipedia.org/wiki/List_of_Moscow_Metro_stations)

> *A pattern is essentially an arrangement. It is characterized by the order of the elements of which it is made rather than by the intrinsic nature of these elements.*
> *Norbert Wiener*

Images made by sets of points as Fig. 2 play a fundamental role in diverse scientific areas as bioecology and complexity theory that investigate the dynamic behavior of swarms, animal herds, bird flocks, and fish schools (Ballerini et al. 2008; Gueron et al. 1996; Reynolds 1987; Bassett et al. 1958).

In particular, the paper of Reynolds interestingly poses the accent on the simulation (as an elaboration of a particle system) of the path of individual birds. It was shown that each bird, represented by a point, moves according to its local perception of the dynamic environment, driven by both the links the bird itself establishes with some of the neighbors (the nearest ones but not only), and by some unknown data collection mechanism (Reynolds 1987). This same process occurs at all sizes, from bacteria to whales, regardless of whether the members work in synchrony or not (Buck and Buck 1968). The paper of Ballerini focuses the analysis on the nature of the interaction between birds. Contrary to a general *a priori* assumption, local or non-local interactions are not based on a metric range or on visual or auditory sensing, but rather on a topological distance. Each bird, represented by a point, can manage a maximum number of connections (information channels), and therefore it proceeds to a selection of neighbors not necessarily based on distance commonly understood. Thus the importance of the analysis of the type of connection is evident. The topological distance substantially poses the accent on the interpretation that the bird, as observer internal to the flock, gives about the data it manages to collect. It would be extremely interesting to compare the nature of the links built by the components of the flock with respect to those established by an external observer of the flock (predator or researcher) (Ballerini et al. 2008).

The study of the interactions applied to animal aggregation, based on images of sets of points (Fig. 1 in Ballerini et al. 2008), can contribute to the understanding of how an external observer selects points from a set and establishes connections between them.

This can be seen taking as an example an image based on a white board spotted with a certain number of points, each with different forms and sizes. The choice of the set of points derives from the intuition of Kandinsky who underlines the power of the point, able to influence the perception of the surface/canvas to which it belongs. The ensemble of identical or nearly identical objects, interacting directly or indirectly, gives rise to the identification of patterns and structures by the observer. More than 200 years of research have not been enough to explain how many different perceptual interpretations can result from one image. The human mind is somewhat strangely satisfied when able to recognize periodicities; in addition our perceptual system manipulates incomplete data; this is why the onset of structures could be biased or affected by ambiguities (Kornmeier and Bach 2009; Zeki 2004). Another important preliminary observation is that when an image is the object of attention, the observer voluntarily or involuntarily reads and re-reads the image and looks for and creates patterns (Glass 1979). In this way from an ordered or disordered set of points, representative or characteristic figures emerge; again, the concept of order is forced by additional geometrical, aesthetical, or functional information. It is notable to remember that the only property of the points in the image is that they belong to the canvas/blackboard and any other property related to their interactions is a characteristic of the observer that operates a selection creating subsets in the set of points.

> *The ideas of common sense pre-exists...in the soul or the human mind: they are the fruit of imagination, which is applied to the data of the senses integrating them and giving them shape.*
>
> *Ernst Cassirer*

When overlaying similar periodic patterns or plotting series of curves on a computer screen, interference phenomena can occur, and moiré patterns can be generated (the adjective moiré characterizes a type of silk textile with a ripple appearance). Random dot patterns together with random dot moiré patterns provide a key to the understanding of the real world based on what the eye usually sees. This is a target important for both artists and scientists, and since all visual phenomena up to the point where they register on the retina are in the domain of optics, art becomes an active branch of applied optics (Oster 1965).

An element of relevant consideration is the role of image perseveration, that is to say the persistence of images on the retina as an after-image. Due to involuntary and small eye movements, the after-image inevitably overlaps with the real image during a further operation of reading (Sarcone and Waeber 2013; Kinsbourne and Warrington 1963). This results in a moiré effect and can happen in both pathological and non-pathological forms. Especially in the case of images where the brain does not agree with what the eyes see, the action of seeing and looking at an image is based on a multiplicity of processes which investigation requires diverse disciplines and for sure psychophysical studies in conjunction with physiological considerations.

The random dot moiré patterns is not an easy task to measure or describe.
Leon Glass

Photoobjects are elements "made of at least two layers, each presenting an image"; we will take as an example a photoobject composed of a white board spotted with a certain number of points and a transparency printed with the same identical image. The observer must superpose and move one layer with respect to the other. In this way the number of points doubles, and structures and patterns emerge (Prete 2004). If the observer performs a small angle rotation, novel evident ring-like structures that cover the whole area of the photoobject appear. If the rotation is at large angles these global ring-like structures, which sometimes appear as spirals, disappear. The operation of multiplication and mutual rototranslation of the two layers of a photoobject is a good illustration of the appearance of collective or emergent properties.

The appearance of a rotational symmetry in the structure can be described by a phase ϕ that together with the number N of objects creates a couple of "observables" that can give a complete description of the dynamics of the system. The two variables N and ϕ are canonically conjugated and, according to the usual formalism of quantum physics, do not commute leading to uncertainty (a feature not present if a classical physics approach is followed). As a consequence, this implies that when one observes the presence of a phase ϕ, it is no longer possible to know exactly the number of points N, i.e., some points of the set are unused and therefore can be classified as invisible (Susskind and Glogower 1964). This operation of selection appears to have links with the metric of the viewed set of points, along the lines discussed in the paper by Ballerini (Ballerini et al. 2008).

An art installation based on a photoobject similar to the one described so far is visible in an untitled work by the Italian artist Pistoletto, who performed, among others, art experiments on moiré patterns (installation visible at *Città dell'Arte* in Biella, Italy). The observer's analysis of the superposition and rotation of two identical images of a transparency spotted with a certain number of points is a useful tool to identify the basic steps of the emergence of patterns: image acquisition, image perseveration on the retina, second image acquisition (after a slight movement, voluntary or involuntary, of head, lids, or eyes), and overlap of the two images. In the case of Pistoletto's installation however, the appearance of patterns cannot rely on retinal perseveration due to the fact that perseveration times are too short, as far as the mechanism of refresh generated by the eye blinking is considered. Of course this exclusion mechanism does not hold if perseverance times are longer than 1 s as in pathological cases. In the case of Pistoletto's installation the overlap is present in some points of the eye-brain path where some recording mechanism is active and acts as a memory. This hypothesis suggests that a similar mechanism is active in the brain of a person observing images while performing a re-reading operation (this time a voluntary one): a bias is then generated by the first reading and results in the presence of artificial structures or patterns. The artist would therefore be right to describe the process as an interaction between two layers of the photoobject (real or generated by reading and re-reading operations supported by memory and visual persistence) built and temporally residing in the brain, responsible for the interference of the two frames.

The freedom of interpretation of a set of points and the contemporary skill to make structures emerge is illustrated in Fig. 3. The reading of the picture emphasizes the ability of the observer to introduce differences in a set of actually identical particles, attributing to some of them a privileged role.

The reproduction of Kandinsky's "Diagram 4—Point—Horizontal-Vertical-Diagonal point pattern for a free line construction" (Kandinsky 1947) in Fig. 3a, b evokes an illustration by Bruno (Division of the corner n. 52) (Bruno 1591) visible in Fig. 3e, f, which is accompanied by an interesting story with lots of unexplored aspects potentially useful for our approach. Of this image two versions exist: the original version of 1591 and a more modern one, called "national" (1879–1891). In the latter, the diagrams have been transformed into a sober rationalized geometric style, where everything that seemed superfluous to the curators was abolished under the assumption that the decorations were meaningless and that Bruno's figures were "wrong" (Bruno 2015).

This rationalization is similar to the operation that a researcher does in an experiment when distinguishing fundamental characteristics from secondary ones; if this criterion is applied with knowledge of the facts it can lead to progress; if it is applied in a sort of random choice, the outcome will quite certainly be tied to a kind of Procrustes' bed. Unfortunately, this second option occurs more often than normally thought and most of the times it is an involuntary process. For example, it commonly happens with photographs where different objects and a mirror are present: people are used to give more (and sometimes total) attention to the objects they see dislocated in the photograph rather than to the reflections of the same objects in the

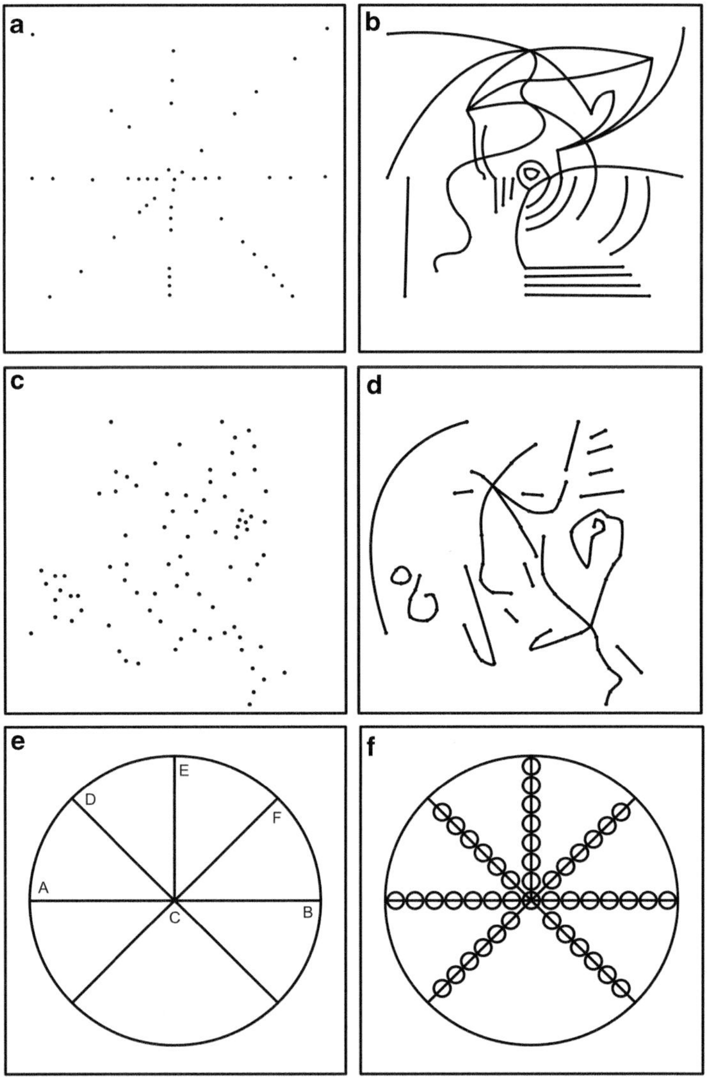

Fig. 3 Examples of possible visualizations associated to the multiplicity of choices that the observer can make. (**a**, **b**) Authors' reproduction of Kandinsky's "Diagram 4—Point—Horizontal-Vertical-Diagonal point pattern for a free line construction" from *Point and Line to Plane* (Kandinsky 1947). (**c**) Barycenters of vesicle distribution in *Staphylococcus aureus* TEM image, without any reference to their size or nanoparticle content, as seen in Fig. 1c in Erega et al. (2017). (**d**) Authors' imagination at work in privileging a subset of dots in (**c**) and drawing links. (**e**, **f**) Geometrical abstract elaborations of a figure from Bruno's *De triplici minimo et mensura* (Bruno 1591, p. 105). This figure with its geometrical significance presents similarities with Kandinsky's point pattern in (**a**), though enriched with symbols and details of somewhat obscure interpretation. (**e**) Authors' elaboration on the basis of the "national" version (1879–1891) of the figure with only circumference, rays, and internal letters. (**f**) Authors' elaboration with rays, circles, and no letters. Both elaborations result from the spoiling of the not easily decipherable signs in the original picture: in this way the path from an image to a pure geometric representation is the reverse of the process illustrated by Kandinsky in the transition from (**a**) to (**b**)

mirror. Moreover people tend to "discount mirror information even when the mirror provides information not otherwise present in the scene," for instance the reflection of objects that are out of frame (Sareen et al. 2015).

Sometimes it is the frame itself that contributes to give some ideas for the interpretation of an image, steering the observer's mind; as widely discussed by Kandinsky in his essay, the frame has great relevance in the perception of the image (Kandinsky 1947).

The availability of several arbitrary choices leads us again to the initial decision of recurring to schemes of sets of points in which established paths and connections will be visually presented in the following figures (Figs. 4, 5, 6, 7, and 8).

The decentralization underlines the existence of a multiplicity of centers, introducing a hierarchy in the set of points that otherwise would be completely identical;

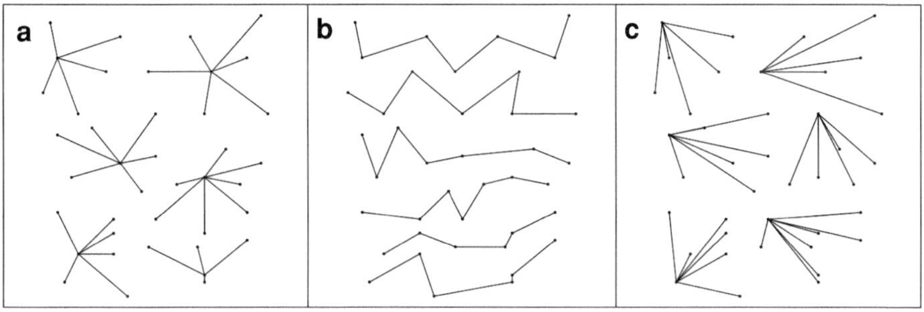

Fig. 4 Multicentric visualizations of the set of points of Fig. 2a corresponding to the different selection of centers, under the arbitrary constraint to maintain the same number of centers. (**a**) Decentralized centrosymmetric structures. (**b**) Decentralized layered structures. (**c**) Decentralized eccentric structures

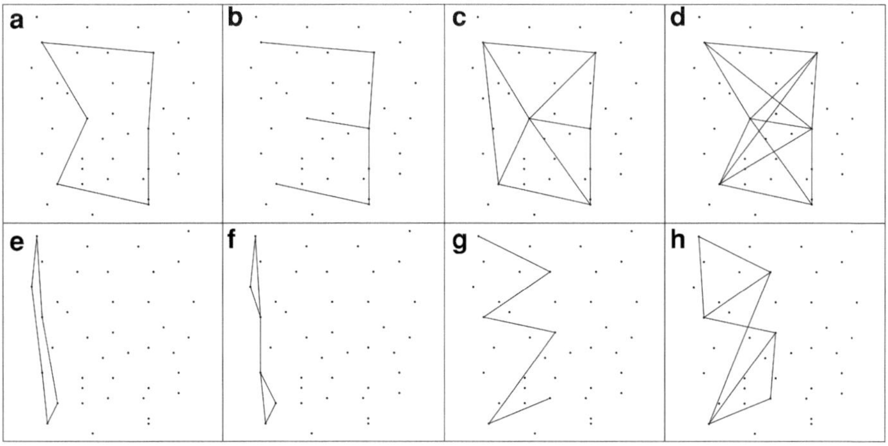

Fig. 5 Decentralization and emerging patterns as a consequence of a multiplicity of centers. (**a–d**) Patterns deriving from the connection of the centers selected in Fig. 4a; (**e–f**) Patterns deriving from the connection of the centers selected in Fig. 4b; (**g–h**) Patterns deriving from the connection of the centers selected in Fig. 4c

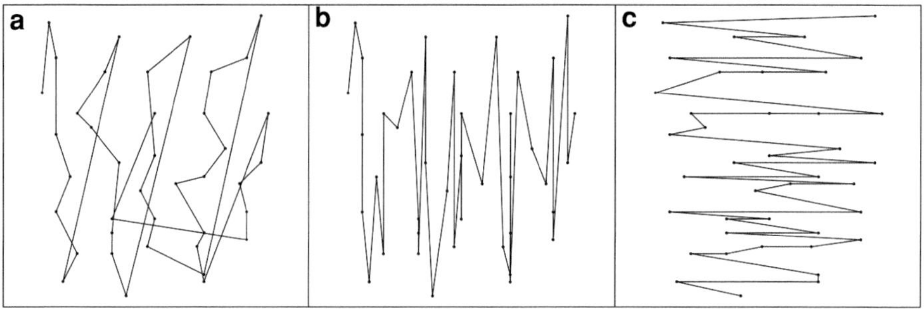

Fig. 6 Importance of the order of data supply in their display and interpretation. (**a**) Automatic uptake of the coordinates. (**b**) Coordinates ordered according to increasing values of abscissae. (**c**) Coordinates ordered according to increasing values of ordinates

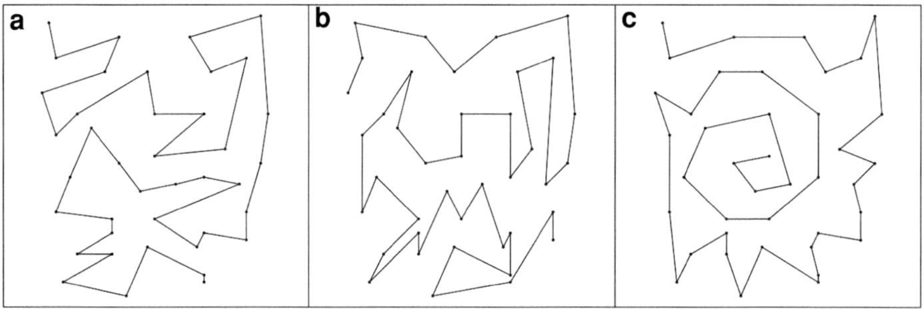

Fig. 7 Importance of the starting point arbitrarily chosen by the observer in the reading of data. Graphs made starting from different points, under the constraint that the line never crosses itself. (**a**) Starting point at top left. (**b**) Starting point at bottom right. (**c**) Starting point in the center

therefore some of the points gain a fundamental role (the centers) and the others keep a subsidiary one. The purely visual partition of the original set of points (Fig. 2a), due to the organizational hierarchy in two subsets, introduces an ambiguity since different readings of the same set of points give rise to different images or pictures (Figs. 4 and 5).

Both Figs. 4 and 5 lead us to take into account the "category theory" where mathematical structures are formalized in terms of graphs, nodes, and arrows (respectively representing concepts, objects, and arrows) so that one does not worry about the properties of the objects investigated, but considers only the interactions and relations to each other; the "category theory" has several applications in different fields from science, to economics and even music (Juniper 2020; Marjanovic 2017; Eilenberg and MacLane 1945). The definition of categories provides the very basics of categorical algebra, characterized by duality, meaning that every definition and theorem has its own reciprocal obtained by "reversing all the arrows." This duality, which is transparent at a theoretical level, is often nebulous in practice and can point to surprising behaviors and relationships (Baez and Stay 2010). The parallelism with Penrose's and Feynman's diagrams is evident (Wright

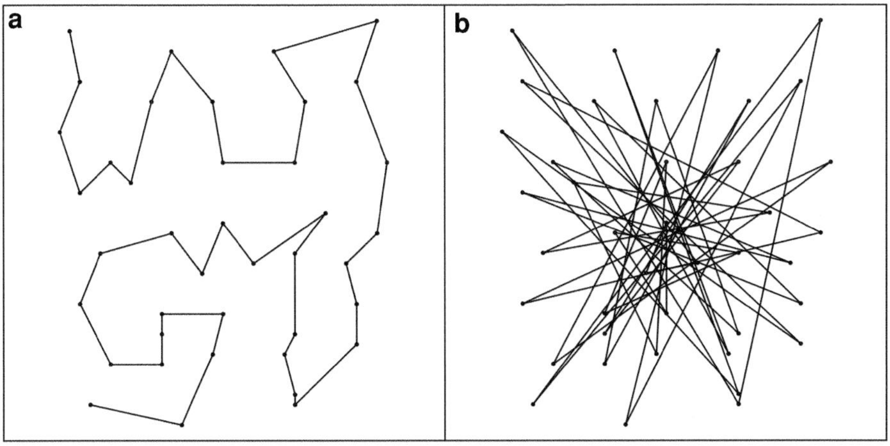

Fig. 8 Importance of short- and long-range connections in the filling of the space. (**a**) Short-range connections made between the nearest neighbors, with only two connections per point available and the constraint to not crossing the links. The short-range interactions give rise to an open structure. (**b**) Long-range connections made between the most distant nodes, with only two connections per point available. The long-range interactions give rise to a closed structure. (**a**) and (**b**) share the same starting point

2013; Devine Thomas 2009), as well as the similarities with the approach we adopted to describe the neverending story. The mechanism of category theory can end up in the emergence of structures and in the discovery of unknown messengers of interaction, as it appears for instance in the investigation via a physical language of the basic forces of nature. A root of this diagrammatic strategy can be found in Bruno, where a particularly interesting approach to the connection with geometry is undertaken through thirty "seals," which have both a mnemonic and an inventive meaning. The images, schemes, diagrams, formal notations, and formal graphics that Bruno called seals are not made up of *phantasmata* but of *signa*: figures that are abstract not only for their content but, not originating from sensations, also for their tangible material. The seals are schematic, topological descriptions of the connections between the notions-images placed in the nodes and represent the possibility of giving a new structure to the parts of a whole, forming a different entity (Bruno 2015). This approach promotes the exploration of the possibilities of rearranging the order of links among the elements. This happens for instance in the Cubist pictorial world where it is possible to look at reality by breaking it down into tiny parts and then reconstructing it on the canvas according to particular rules; the multiplication of points of view, deriving from the partition of the pictorial surface, results in the dissolution of the traditional perspective, where each small surface records a different standpoint so that the observer can make a sort of virtual itinerary over space and time.

The interpretation of an image or set of points can be influenced also by the sequence of data feeding. This aspect that underlies visualization activities in a competition between serial and parallel processes is present in the human observation, but is more evident in automatic image analysis, a fundamental step when a high number of images need to be analyzed.

As an example we reported in Fig. 6 different outputs generated by changes in the order of data acquisition obtained with the software Graph4Win. Choices were extremely arbitrary and only weakly deriving from the nature of the links between points. For simplicity we imposed some constraints on the number of incoming and outgoing lines from each node accepting a maximum of two. Since the acquisition of electron microscopy image data does not always involve an absolute spatial reference, the comparison of Fig. 6b, c is significant.

Similarly, Fig. 7 shows how the selection of the starting point can influence the structure identification in a certain order of data.

From models such as those of electrical dipoles we know that short-range interactions take place, but it is necessary to take into account also long-range ones that can have a great role in dictating collective dynamics and structures. Long-range interactions were exemplified in the works of Alexander S. Davydov and Herbert Fröhlich (Davydov 1985; Fröhlich 1977). A gas of electrons presents dynamics and interactions dictated by usual Coulombian repulsion, but attractive forces in the presence of a medium can also be at work as in the case of a superconductor characterized by the formation of Cooper pairs. Both the cases are extremely interesting for us since nanoparticles can present complex surface electrical distribution.

Figure 8 represents the result of image reading when points are connected by short-range or long-range links. This is a case very similar to what happens when a traveler, looking to the metro map of an unknown town, focuses on the overall path of one metro line, its orientation, its length, and so on (long-range connection), or on the contrary, pays attention just to the small segment connecting two nearby stations (short-range connection) to evaluate their distance and consequently decide whether to save money and walk.

Looking at Fig. 8a, b, taking into account their different relation with the environment and the diverse ability to fill the space according to the short- or long-range interaction, it is evident that the segments imaged are the representation of the fractional dimension for a curve inside a plane represented by a number between 1 and 2 which somehow expresses the degree of complexity of the curve itself (Hausdorff's dimension) (Mandelbrot 1989).

As it has been told so far, images are composed of a combination of points, and of the correlating lines that contribute to the formation of various possible arrangements. In literature the generation of lines from points is well illustrated in Calvino's *The Castle of Crossed Destinies* with the tarot arrangement. In the story the tarot has the feature of a point that, connected to other tarots, describes and builds up one or more stories, the same mechanism for which the connection of points creates one or more images (Calvino 1977). This process has been scientifically discussed by artists as Kandinsky and Kazimir S. Malevich (Fig. 9) whose studies led to the identification of the point as the founding element of every visual representation and image formation, expression, and interpretation (Kandinsky 1947; Malevich 1915). To support their theories Kandinsky and Malevich produced highly detailed

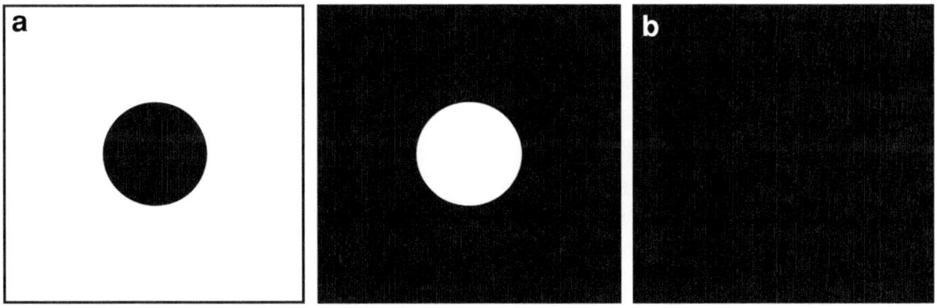

Fig. 9 (**a**) Authors' reproduction of Kandinsky's "Diagram 5—Point—The black and white points as elementary color values" from *Point and Line to Plane* (Kandinsky 1947). (**b**) Authors' reproduction of Malevich's *Black Square*—State Tretyakov Gallery, Moscow, Russian Federation. Kandinsky's points consist of a black round dot on a white surface delimitated by a black square, and of a white round dot on a black square. Both the points and the black square are mediated by a surface (explicated only by Kandinsky and just implied by Malevich) and have the role of *tesserae* that create a mosaic-like image when placed on a canvas

illustrations and tables becoming the promoters of Constructivism (Bannister 2012). The correlation of the works of a literate and two painters makes more evident the necessary dialectics not only between the two macro areas of culture, art and science, but also among adjacent fields as art and literature.

The arrangement of points was not only investigated in the art and literature fields but also in the scientific area by Dmitri I. Mendeleev. In 1866 Mendeleev began writing *Principles of Chemistry* with the aim of putting order and systematizing the information of the 63 chemical elements known at the time. He played the magician's game: he took 63 cards where he wrote the detailed characteristics of each element and then proceeded to order the cards in all possible ways trying to see whether some pattern or some regularity emerged. As he shuffled the cards, he realized that when organized in a certain order the cards showed the appearance of periodicity in the elements' chemical properties. In this way he built the table that nowadays is called after him (Mendeleev 1871). The periodic table of elements encodes the language of matter, and as the combination of the 26 letters of the alphabet gives rise to all the words of our vocabulary, so the combination of simple elements gives rise to all the substances present in nature (Mussardo and Polizzi 2019). As Calvino used the cards to build combinations and tell stories, so Mendeleev arranged and rearranged the cards of the elements in many different ways until he made a pattern emerge; however, differently from the writer, the scientist did not arbitrarily combine elements but studied them to find order and patterns, eventually leaving to nature the combinatorial activity of building substances. Humanities and scientific activities build interpretative models of reality that are subjected to continuous evolution and mutation, destined to be always questioned; the two areas share the use of narrative forms, of which the maximal expression is the thought experiment, or *Gedankenexperiment*, which has the same conceptual value as a laboratory experiment or reality manipulation (Fabbri 2002).

As previously seen, the interpretation of points and lines is biased by the environment in which they are embedded since both the surface and the boundary play an important role. Actually, for Kandinsky the surface is not just part of a canvas, but it is endowed with a duplicity of manifestations in terms of color (black and white, or absence and presence of light), the contrast of which creates a line, the boundary, that is fundamental for the definition of the point and its characteristics. The analysis of the boundary formation is a key point in Kandinsky's theory; at the same time, it forms a loop since a line is necessary to define a point that in turn is the basic element for the definition of the line (Kandinsky 1947).

The juxtaposition of black and white recalls Bruno's image formation where the visualization of concepts implies the identification of image details that in a verbal definition may remain undefined or uncertain (Bruno 2001). For instance the line, traditionally defined via the verbal paradox of an object with some length and without thickness, may and must be visualized by the creation of a boundary between a "clear" zone and an obscure one. This discussion points out the activity of Bruno as both a painter and an investigator of image formation, underlying the mechanisms proper of the *chiaroscuro* technique based on contrast between light and dark that leads to the achievement of a sense of volume in modeling bidimensional objects and figures. This technique had a strong propulsion in Caravaggio, a contemporary of Bruno. Both Bruno and Caravaggio, who lived in the same cultural atmosphere having shared the ambient of the church of Rome in the last years of the sixteenth century, were outstanding solitary, nonconformist, and rebel figures for their art and philosophy and had in common the research of the formation of images based on the pivotal role of mirror and shadow (Panzera 2011).

The dialog between science and abstract art had the aim to establish a science of the art through the primary elements of the form: point, line, and plane, the key elements that physically support and condition, through the material and surface's properties, the existence of the opera.

Kandinsky, like a lot of other artists, was an observer of the world who managed to fix on canvas his observations; he was not just a passive observer but a transformer because he went beyond the study of the fundamental elements of image raising importance on the role of the observer's subjective view, influence that affected the paintings and their relation with the actual world. Along this path he gave scientific form to the inverse or curvilinear perspective typical of icons; such perspective is related to non-Euclidean geometry, which characteristic is that in the pictorial space straight lines are represented as curved ones. This distorted perspective is strictly connected to interpretation: knowledge extraction implies the *anamorphosis* of reality and acts as human vision in an approach to explain phenomena normally scrutinized through hard sciences' lens (Borkhardt 2017).

Kandinsky's adoption of non-Euclidean geometry, already largely used in a metaphorical, non-scientific sense of the term, demonstrates the interest of the artist in the connection of different artistic forms in the two cultures (Antonova 2020; Short 2010). Kandinsky's pursuit of a common root in various arts was inspired by a work by Pushkin regarding an image known since the Middle Ages under the name of Salt-Point and present in Bruno's essay "*De triplici minimo et mensura*" (Bruno

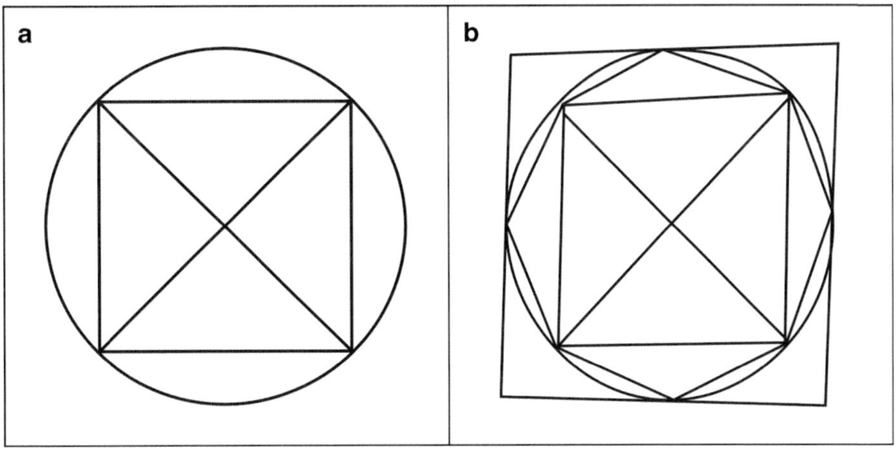

Fig. 10 (**a**) Authors' reproduction of Pushkin's "Triangle and square in circle, as basic source for numerals: Arabic and Roman" from Kandinsky's *Point and Line to Plane* (Kandinsky 1947). This figure was also known as Salt-Point. (**b**) Geometrical abstraction of Bruno's *Antiphontis tetragonismus* from "*De triplici minimo et mensura*" (Bruno 1591). The irregularities present in the original illustration by Bruno have been respected and reported by the authors in this reproduction

1591) under the name "*Antiphontis tetragonismus*" (Fig. 10). Pushkin affirmed that the geometrical figure, which includes a triangle, a square, and a circle, shows the products of a moving point and in its more abstract form can represent the roots of "two numerical systems" (Pushkin 1955). The elements present in the Salt-Point figure are still of great interest in current semiotic and philosophical debates (Marconi 2020).

In Kandinsky's works the interplay between art and science is continuously in evolution, anticipating a lot of the problems and solutions regarding the digital production of images and object recognition in computer vision, where monitor and pixels take the role of the artist's canvas and points, respectively.

Trying to gain experience from all that has been said so far about the interpretation of electron microscopy images, it can be useful to recur to the graphics previously presented in this chapter that are a good example of data collection, interpretation, and extrapolation. A set of points collocated on a plane are analyzed, interpreted, and "frozen" by an operator in an image that will be subjected to further analyses and interpretations (re-readings) by a researcher (sometimes the operator and researcher are the same person). These passages are similar to what Kandinsky did in his activity and his research of the form, where the final meaning of the opera is only suggested by the artist, but actually interpreted by the observer, in a kind of mutual relation of both the artist and the observer.

The set of points that in an abstract manner is the result of the extraction of information from an electron microscopy image can be seen as an homogeneous and uniform state from which patterns, sometimes time-dependent and ambiguous, can emerge as described with a mathematical approach to the concept of shape in the works by Turing and René Thom (Thom 2018; Turing 1990). The morphogenesis theory is a basic model in theoretical biology that found connections and developments in different fields: from dissipative structure, hypercycles, and weak condensation to synergetics and pattern formation, related to spontaneous, self-organized emergence of structures in material and immaterial systems generally composed of many individual identical parts, even though the relation of sameness and identity opens to further discussions because it can be a subjective issue (Haken 1979, 1978; Eigen and Schuster 1978; Fröhlich 1977; Nicolis and Prigogine 1977).

In his representations Kandinsky introduced the division in elemental components as fundamental entities that required to be analyzed and organized in sets. Towards the end of 1800 other artistic movements as Italian Divisionism and French Pointillism proposed a different technique to build an image by placing small dots of paint alongside one another on a canvas. Though the eye can jump from one dot to the next perceiving them as distinct units, the closeness of the points creates an optical fusion, a kind of phase transition in the visual perception that makes the eye able to see as a unique image the resulting combination (Molholt 2011). The two art movements research and explore nature as does science. The link between the two cultures is maintained by the presence and contamination of science in art and literature. What we aim to do here is walk the reverse path, using science to investigate artistic and literal processes, *de facto* bringing literature and art into science. Following this approach, the visual experiments of Divisionism and Pointillism can be associated in the physical world for instance to the Josephson's junctions since they are structures born by the mere closeness of two superconductors, or for instance to the two-level description of the atom, fundamental in laser modeling, where the atom's state (wave function) presents itself to the observer in the form of a suitable combination of the two (eigen)states, each one associated to the occupation of one of the two levels. Along the same approach also in Kandinsky's "Diagram 5—Point—The black and white points as elementary color values" (Fig. 9a), the white color of the square's interior dictates the black color of the point and vice versa; therefore each point, just as an electron, is endowed with a twofold degree of freedom (spin ½) representing the possibility of the two choices. Following this quantum mechanical vision, when we take into account a series of identical points close enough to one another and perform a statistical analysis studying their collective behavior, we can say that in Kandinsky's work every point has the properties of a fermion (technically a particle with a half integer spin or angular momentum), so that each point bounded by the square behaves like an electron. Assigning a fermionic behavior to these elementary units means that these units obey the Pauli exclusion principle and that the units cannot condensate (or fuse together, occupying the same position in space), i.e., they are somewhat repelling each other. On the contrary if two elementary units are close enough, they can achieve the status of a single body and turn the ensemble of fermionic units into an ensemble of bosons (spin 1 or integer spin or angular momentum), giving rise in physical terms to a Bose-Einstein condensate equivalent to the Divisionists' optical fusion.

The dualism of the point as shown in Kandinsky's "Diagram 5—Point—The black and white points as elementary color values" (Fig. 9a) can be also associated to an object and its shadow or to the presence and absence of an object. This dualism is very popular in physics (for instance in semiconductor physics when talking about electrons and holes) and obvious in the process of photograph formation, when associated to black and white pixels and consequently to the couples light and dark, matter and void. A deeper contribution about the analysis of this duality is given once again by literature, when Calvino's Mr. Palomar that "always hopes that silence contains something more than what language can say" recalls some verses from an ancient Chinese text: "Thirty spokes converge on the hub, but it is the void at the center of the wheel that makes the cart move." Mr. Palomar gives also a further lesson about the importance of the void, a key passage to be kept in mind when analyzing electron microscopy images: "interpreting therefore means giving a peculiar shape to the void" (Zublena 2002; Calvino 1985).

Through the course of this book we pointed out the intricate relationship existing between humanities and science, with particular attention to images. Even those obtained with the most cutting-edge technologies must pass under the complex lens of human vision, which in turn is at the base of the reading and interpretative processes ruled by the mind.

Scientists' descriptions of concepts, phenomena, or processes enclose all the inflexions of the word "image," emphasizing the several meanings that it can have according to the context and the general sense of the sentence. As an example Bruno based on images his mnemonic techniques, Heinrich R. Hertz stated that he visualized his "conception of things" through images, Ludwig E. Boltzmann talked of images as "mental pictures of phenomena," and Einstein said that "a man ... may form some pictures ... of the things he observes" (Cellucci 2015). Visualization is also the key point of the word *anschaulich*, a German term that can be translated either as visualizable and appealing to graphical imagination or as intuitive. Heisenberg introduced this ambiguous term to describe his "quantum uncertainty" to those who felt his theory too abstract (Werner and Farrelly 2019). Finally, it must be recalled that in physics diagrams take a central role, as both an intuitive aid in understanding the world and a powerful tool for performing calculations as shown by Penrose and Feynman (Selby 2014; Wright 2013).

The connections that images have with life and with disciplines belonging to both the two cultures show that a communication between science, literature, and art is so strongly required that without one, the others could not develop, thus compromising the evolution of the socio-cultural environment. However, the growing specialization of knowledge that demands always more in-depth analyses paradoxically seems to lead to the alienation of knowledge itself, where even a very valued work if taken out of context can be understood by just a few people. The continuous division of subjects in subcategories, together with the missing communication between scholars, contributes to the disintegration of knowledge, making it less available to the public (Perrone Capano 2011).

In the modern society where common and basic knowledge is within arm's reach in most parts of the world, mere notions without personal intuitions and interpretations will not give any more meaning to knowledge. The key to restoring a joint conscious culture is keeping the dialogue open between the knowledge and the sensitivity that every scientist, observer, reader, or artist has within.

References

Antonova C. Non-Euclidean geometry in Russian art history: on a little-known application of a scientific theory. Leonardo. 2020;53(3):293–8. https://doi.org/10.1162/leon_a_01786.

Baez J, Stay M. Physics, topology, logic and computation: a Rosetta Stone. In: Coecke B, editor. New structures for physics. Berlin: Springer; 2010. p. 95–172. https://doi.org/10.1007/978-3-642-12821-9_2.

Ballerini M, Cabibbo N, Candelier R, Cavagna A, Cisbani E, Giardina I, Lecomte V, Orlandi A, Parisi G, Procaccini A, Viale M, Zdravkovic V. Interaction ruling animal collective behavior depends on topological rather than metric distance: evidence from a field study. Proc Natl Acad Sci. 2008;105(4):1232–7. https://doi.org/10.1073/pnas.0711437105.

Bannister F. Revolutions in time, space, and art: Russian constructivism. UNIversitas. 2012;7(1):article 5. Retrievable at https://scholarworks.uni.edu/universitas/vol7/iss1/5

Bassett GA, Menter JW, Pashley DW. Moiré patterns on electron micrographs, and their application to the study of dislocations in metals. Proc R Soc London A Math Phys Sci. 1958;246(1246):345–68.

Bauman Z. La decadenza degli intellettuali. Da legislatori a interpreti. Torino: Bollati Boringhieri; 2007.

Borkhardt S. 'Russian Messiah': on the spiritual in the reception of Vasily Kandinsky's art in Germany, c. 1910–1937. In: Hardiman L, Kozicharow N, editors. Modernism and the spiritual in Russian art. Cambridge: Open Book; 2017.

Bresciani Califano M. Italo Calvino e la scienza. In: Bresciani Califano M, editor. Piccole zone di simmetria: scrittori del Novecento. Firenze: Firenze University Press; 2011.

Bruno G. *De triplici minimo et mensura*. 1591.

Bruno G. *Corpus iconographicum*. Le incisioni nelle opere a stampa. London: Adelphi; 2001.

Bruno G. Il Sigillo dei sigilli. I diagrammi ermetici. Giordano. Bruno filosofo e pittore. Milano: Mimesis; 2015.

Buck J, Buck E. Mechanism of rhythmic synchronous flashing of fireflies: fireflies of Southeast Asia may use anticipatory time-measuring in synchronizing their flashing. Science. 1968;159(3821):1319–27. https://doi.org/10.1126/science.159.3821.1319.

Calvino I. The castle of crossed destinies. London: Martin Secker & Warburg; 1977.

Calvino I. Mr. Palomar. London: Martin Secker & Warburg; 1985.

Cannon J. Calvino's latest challenge to the labyrinth: a reading of Palomar. Italica. 1985;62(3):189–200. https://doi.org/10.2307/479029.

Cellucci C. Rethinking knowledge. Metaphilosophy. 2015;46(2):213–34. https://www.jstor.org/stable/26602301

Davydov AS. Solitons in molecular systems. Dordrecht: Springer; 1985. https://doi.org/10.1007/978-94-017-3025-9.

Devine Thomas K. Feynman diagrams and the evolution of particle physics. 2009. Retrievable at https://www.ias.edu/ideas/2009/arkani-hamed-oconnell-feynman-diagrams

Eigen M, Schuster P. The hypercycle. Naturwissenschaften. 1978;65:7–41. https://doi.org/10.1007/BF00420631.

Eilenberg S, MacLane S. General theory of natural equivalences. Trans Am Math Soc. 1945;58(2):231–94. https://doi.org/10.2307/1990284.

Erega A, Curia R, Milani M. Interactions between nanoparticles and bacterial structures as seen through software enhanced analysis of electron microscopy images. In: Méndez-Vilas A, editor. Microscopy and imaging science: practical approaches to applied research and education. Formatex Research Center; 2017. p. 587–94.

Fabbri P. Scienza, quante storie! 2002. Retrievable at https://www.paolofabbri.it/interviste/scienza_quante_storie/

Fröhlich H. Long-range coherence in biological systems. Riv Nuovo Cimento (1971–1977). 1977;7(3):399–418.

Glass L. Physiological mechanisms for the perception of random dot Moiré patterns. In: Haken H, editor. Pattern formation by dynamic systems and pattern recognition. Berlin: Springer; 1979. https://doi.org/10.1007/978-3-642-67480-8_12.

Gueron S, Levin SA, Rubenstein DI. The dynamics of herds: from individuals to aggregations. J Theor Biol. 1996;182(1):85–98. https://doi.org/10.1006/jtbi.1996.0144.

Haken H. Synergetics. An introduction. Nonequilibrium phase transitions and self-organization in physics, chemistry and biology. Berlin: Springer; 1978. https://doi.org/10.1007/978-3-642-96469-5.

Haken H, editor. Pattern formation by dynamic systems and pattern recognition. Berlin: Springer; 1979. https://doi.org/10.1007/978-3-642-67480-8.

Hardin R. Trust and trustworthiness. New York: Russell Sage Foundation; 2002.

Huotari J, Helenius A. Endosome maturation. EMBO J. 2011;30:3481–500. https://doi.org/10.1038/emboj.2011.286.

Juniper AJ. Semantic technologies for disaster management: network models and methods of diagrammatic reasoning. In: Chaiechi T, editor. Economic effects of natural disasters. London: Academic Press; 2020. p. 179–90. https://doi.org/10.1016/B978-0-12-817465-4.00012-1.

Kandinsky VV. Point and line to plane. New York: The Solomon R Guggenheim Foundation; 1947.

Kinsbourne M, Warrington EK. A study of visual perseveration. J Neurol Neurosurg Psychiatry. 1963;26(5):468–75. https://doi.org/10.1136/jnnp.26.5.468.

Kornmeier J, Bach M. Object perception: when our brain is impressed but we do not notice it. J Vis. 2009;9:7. https://doi.org/10.1167/9.1.7.

Malevich DI. Black square. Moscow: State Tretyakov Gallery; 1915.

Mandelbrot BB. La Geometria della Natura. Sulla teoria dei frattali. Roma: Theoria; 1989.

Marconi V. La linea, il triangolo e il quadrato. La natura relazionale di segno e significato in Aristotele, Peirce e Hjelmslev. PhD Dissertation. 2020.

Marjanovic M. Construction of physical models from category theory. Master thesis in theoretical physics. 2017.

Mendeleev DI. Osnovy khimii [The principles of chemistry]. 1871.

Molholt R. Roman labyrinth mosaics and the experience of motion. Art Bull. 2011;93(3):287–303. https://doi.org/10.1080/00043079.2011.10786009.

Mussardo G, Polizzi G. L'infinita scienza di Leopardi. Trieste: Scienza Express; 2019.

Nicolis G, Prigogine I. Self-organization in non-equilibrium systems. From dissipative structures to order through fluctuations. Chichester: Wiley; 1977.

Oster G. Optical art. Appl Opt. 1965;4(11):1359–69.

Panzera AM. Caravaggio, Giordano Bruno e l'invisibile natura delle cose. Roma: L'Asino d'Oro; 2011.

Perrone Capano L. Premessa. Letteratura e scienza: un intreccio ineludibile. 2011.

Prete N. Doubling the image to face the obscenity of photography. In: Globus GG, Pribram KH, Vitiello G, editors. Brain and being. At the boundary between science, philosophy, language and arts. Philadelphia: John Benjamins; 2004.

Pushkin AS. Works, vol. 16. Annenkov: Petersburg Verlag; 1955.

Reynolds CW. Flocks, herds and schools: a distributed behavioral model. In: Stone MC, editor. SIGGRAPH '87: proceedings of the 14th annual conference on Computer graphics and interactive techniques; 1987. p. 25–34. https://doi.org/10.1145/37401.37406.

Ricci F. Painting with words, writing with pictures: word and image in the work of Italo Calvino. Toronto: University of Toronto Press; 2001.

Salamone N. Postmodernità. Quotidiano e orizzonte nella società contemporanea. Roma: Carocci; 1999.
Sarcone GA, Waeber M-J. Impossible folding puzzles and other mathematical paradoxes. Mineola: Dover; 2013.
Sareen P, Ehinger KA, Wolfe JM. Through the looking-glass: objects in the mirror are less real. Psychon Bull Rev. 2015;22:980–6. https://doi.org/10.3758/s13423-014-0761-8.
Sartori G. Homo videns. Televisione e post-pensiero. Roma: Laterza; 2000.
Selby JH. An overview of diagrammatic notation in physics. CQD MRes report. London: Imperial College; 2014.
Short S. The art theory of Wassily Kandinsky, 1909–1928. the quest for synthesis. New York: Peter Lang AG, International Academic Publishers; 2010.
Susskind L, Glogower J. Quantum mechanical phase and time operator. Physics. 1964;1(1):49–61.
Szymura M. Il sapere indifeso in "Palomar" di Italo Calvino. In: Wojtynek-Musik K, Parisi A, Parisi GL, editors. La sfida eraclitiana nella narrativa italiana postmoderna. Katowice: Wydawnictwo Uniwersytetu Śląskiego; 2009. p. 78–101. http://hdl.handle.net/20.500.12128/3893.
Thom R. Structural stability and morphogenesis. An outline of a general theory of models. Boca Raton: CRC Press; 2018.
Turing AM. The chemical basis of morphogenesis. Bull Math Biol. 1990;52:153–97. https://doi.org/10.1007/BF02459572.
Werner RF, Farrelly T. Uncertainty from Heisenberg to today. Found Phys. 2019;49:460–91. https://doi.org/10.1007/s10701-019-00265-z.
Wright S. Trust and trustworthiness. Philosophia. 2010;38:615–27. https://doi.org/10.1007/s11406-009-9218-0.
Wright AS. The origins of Penrose diagrams in physics, art, and the psychology of perception, 1958–62. Endeavour. 2013;37(3):133–9. https://doi.org/10.1016/j.endeavour.2013.02.001.
Zeki S. The neurology of ambiguity. Conscious Cogn. 2004;13(1):173–96. https://doi.org/10.1016/j.concog.2003.10.003.
Zublena P. L'ultimo Calvino tra precisione e disastro. In: Zublena P, editor. L'inquietante simmetria della lingua. Il linguaggio tecnico-scientifico nella narrativa italiana del Novecento. Torino: Edizioni dell'Orso; 2002.

Acknowledgments

The authors are indebted to Alessandro Erega for the time and effort spent in the figures' preparation and editing. We are grateful to Alessandro for his patience and support, and for putting his experience and knowledge to our disposal. The constructive conversations and advice were deeply appreciated.

The authors acknowledge the work of Viaceslav Aleksievich Galkov, director of Нанопромимпорт (Moscow, Russian Federation), for the accurate and thorough collection of contributions for the Memorial Day in honor of Lyubov Vasilievna Didenko (Moscow, April 6, 2016).

The authors wish to thank:

Claudio Savoia (ST Microelectronics, Agrate Brianza, Italy) for a long-lasting friendship that started at the Institute of Physics of the University of Milan (Milan, Italy), and for the days spent in front of FIB and SEM microscopes exploring some of the most unusual samples;

Damjana Drobne (University of Ljubljana, Slovenia—Department of Biology) for her enthusiastic incitement and support for electron microscopy in general, and in particular for the applications of focused ion beam microscopy to investigations in the biological and toxicological fields;

Edmond Turcu (Science and Technology Facilities Council—STFC—Central Laser Facility, Rutherford Appleton Laboratory, Oxford, UK) for his precious suggestions on the potential of FIB microscopy as both a general-purpose investigation tool and a non-conventional one for measurements;

Henri Lezec (National Institute of Standards and Technology—NIST, Gaithersburg, MD, USA) for the open-minded discussions and the investigation of biological samples with FIB microscopy.

A thank you to Marco Barberi for his kind support, for the stimulating and endless conversations rich of beautiful connections between the two cultures, for his availability in sharing much of his knowledge and philosophy of life, and for always sincerely and honestly expressing his thought.

A special thanks to Olivia Charlotte Gilmore for her kind contribution to the text review.

M. Milani et al., *Bacterial Degradation of Organic and Inorganic Materials*,
https://doi.org/10.1007/978-3-031-26949-3

Finally, the authors express their gratitude to the staff at Springer Nature for the kind collaboration, and for the invaluable activity and contribution provided to make this project become real.

Appendix: "Mirror Reflections" in Art, Humanities, and Science

"The normal mirror is a prosthesis that does not deceive."	Eco U, 1997 Authors' translation	Literature
"The mirror tells the truth. It doesn't translate … It speaks the truth in an inhuman extent."	Eco U, 1985	Literature
"Mirrors can lie."	Bacchini F, 1995 Authors' translation	Literature
"People have a naive understanding of how mirrors behave … Our commonsense conceptual and—to a lesser extent—perceptual understanding of reflection is limited and biased."	Hecht H, 2005	Psychology
A mirror "registers what strikes it exactly as it strikes it." The mirror does not interpret objects.	Peters BJ, 2017	Philosophy/ religion
"A mirror can reflect anything except itself."	Pistoletto M, 1977	Art
"The key characteristics of mirror neurons are that their activity is modulated both by action execution and action observation."	Kilner JM, 2013	Science
The theory of reflection can have a physical "view" and a metaphorical interpretation.	Karpatschof B, 2000	Psychology
Mirrors have numerous mystical connotations, and the reflected images are appreciated for their power of revelation[a].		Philosophy/ religion
"*Videmus nunc per speculum et in aenigmate, tunc autem facie ad faciem.*"	Saint Paul, ~55 AD	Philosophy/ religion
"The mirror as means for self-reflection and moral judgment, the mirror as reliable reflection of events, and the mirror as reflection of the nature of a newspaper's readership or market."	Vos T, 2011	Sociology/ journalism
"A social self … might be called the reflected or looking-glass self: … each to each a looking glass—Reflects the other that doth pass."	Rousseau N, 2002	Sociology
"*Imago animi vultus, indices oculi.*"	Cicero MT, 55 BC	Literature
"The eyes are the mirror of the soul and reflect everything that seems to be hidden; and like a mirror, they also reflect the person looking into them."	Coelho P, 2012	Literature

M. Milani et al., *Bacterial Degradation of Organic and Inorganic Materials*,
https://doi.org/10.1007/978-3-031-26949-3

"Behaviour is the mirror in which everyone shows their image." *Translation by Stopp E.* "A man's manners are a mirror in which he shows his portrait." *Translation by Saunders TB.*	von Goethe JW, 1833	Literature
All the people we meet are our reflection. They have been sent to us because, seeing them, we correct our mistakes; and when we do it these people change too.	Pasternak BL	Literature
"Our self-concepts are formed as reflections of the responses and evaluations of others in our environment."	Gecas V, 1983	Psychology/ sociology
"The universe is the mirror in which we can contemplate only what we have learned to know in ourselves."	Calvino I, 1985	Literature
"It is no use to blame the looking-glass if your face is awry."	Gogol NV, 1836	Literature
"The very act of a person seeing himself in a mirror or being represented in a portrait as the center of attention encouraged him to think to himself in a different way."	Mortimer I, 2016	Psychology
"Se reconnaître dans le miroir … Comment le miracle va-t-il s'accomplir? Reconnaître dans la glace le visage d'autrui est relativement facile. L'enfant se retourne et il constate l'identité de l'image avec son modèle. Mais l'image de soi dans le miroir n'a pas de modèle. Elle n'est pas et ne sera jamais comparable à la vision directe de soi-même."	Zazzo R, 1993	Psychology
"The painter's mind should be like a mirror, which transforms itself into the color of the thing it has as its object, and is filled with as many likenesses as there are things placed before it."	Leonardo da Vinci, ~1515	Art
"The ability to recognize oneself in the mirror is often held as evidence of self-awareness."	Huttunen AW, 2017	Psychology
"Some of the earliest explorations … used mirror tests to determine if animals showed evidence of self-awareness, that is, an ability to separate their concepts of their own bodies (self) from the bodies of others."	Breed MD, 2012	Psychology
"Je mis en parallèle les réactions devant le miroir, les réactions aux photographies, les réactions aux images cinématographiques, et aussi bien pour ces trois types de "doubles" les réactions à l'image d'autrui qu'à l'image de soi."	Zazzo R, 1993	Psychology
"When I was a boy, I spent hours and hours in front of the mirror pulling faces … So that I could believe I was a different person, many people of every kind, a host of individuals who one after the other became myself."	Calvino I, 1985	Literature

"This way [of creating a poem] is not difficult. It can be quickly managed anywhere on earth—most quickly if you are prepared to carry a mirror with you wherever you go."	Plato, ~375 BC	Philosophy
Narcissus saw his own reflection in a pond and fell deeply in love with it. The impossibility to concretely reach the reflection and hence have his love reciprocated, led him to desperation and eventually to death.	Ovidius PN, ~8 AD	Literature
Human beings were not allowed to look directly at the divinity without crossing the threshold of the Hereafter, but men could look at the divinity through reflection.	Ovidius PN, ~8 AD	Literature
"Scudo con testa di Medusa." The reflected images of Medusa in Perseus' mirrored shield.	Caravaggio, 1598	Art
The mirror becomes an active character in fairy tales: sometimes it can be magical, some other enchanted and its user is allowed to see on its glass anything wished. "The Beast concealed himself inside his castle, with a magic mirror as his only window to the outside world." "She had a magic mirror into which you could look and find out what was happening and where."	Barbot de Villeneuve GS, 1740 Afanas'ev AN, 1855	Literature
"Art is the magic mirror you make to reflect your invisible dreams in visible pictures. You use a glass mirror to see your face: you use works of art to see your soul."	Shaw GB, 1920	Literature
"Eh, monsieur, un roman est un miroir qui se promène sur une grande route."	Stendhal, 1830	Literature
"Le miroir à deux faces." "Dans le miroir que tu regardais par-dessus l'épaule ..., que voyais-tu exactement?"	Tournier M, 1994	Literature
"Sometimes, at the back of the mirror, behind my reflection, I thought I saw a presence I wasn't quick enough to identify and which immediately hid."	Calvino I, 1985	Literature
Then the cow asked: "What is a mirror?" "It is a hole in the wall," said the cat. "You look in it, and there you see the picture."	Twain M, 1909	Literature
"And on that table stands Mirror with a candle... Make a guess, Svetlana; In clean mirror glass At midnight, no deception You will know your lot."	Zhukovsky VA, 1813	Literature

"The girls, having taken a mirror, some water in a bowl and some barley grains put all this on the floor and let loose chickens. If a chicken goes to the mirror and looks into it the husband will be a dandy; if it drinks the water this portends a drunken husband; if it pecks the grains this means a rich husband."	Ryan WF, 1992	Literature
"Tatyana, in her low-cut gown, Steps out of doors and trains a mirror Upon the moon to bring it nearer."	Pushkin AS, 1833	Literature
"Some have believed that the Moon has some light of its own, but this opinion is false, for they have based it upon that glimmer visible in the middle between the horns of the new Moon … this brightness at such a time being derived from our ocean and the other inland seas, for they are at that time illuminated by the Sun, which is then on the point of setting, in such a way that the sea then performs the same office for the dark side of the Moon as the Moon when at full does for us when the Sun is set."	Leonardo da Vinci, ~1510	Science
The "lunar laser ranging retroreflector array" is a 2-foot-wide panel studded with 100 mirrors (corner-cube reflectors) pointing at Earth [Apollo 11, 1969]. A laser pulse is reflected. "Back on Earth, telescopes intercept the returning pulse, "usually just a single photon"."	NASA Science, 2004 https://science.nasa.gov/science-news/science-at-nasa/2004/21jul_llr	Science
Looking glass: old-fashioned term to indicate an object with reflective properties. According to different external conditions, the reflection can be either partial (hence the observer is allowed to look through the looking glass) or total (as in a classical mirror). Mirror: a polished or smooth surface (as of glass) with a shiny, metal-covered back that reflects light, producing an image of whatever is in front of it. The observer is not allowed to cross the background surface or looking through the mirror, but only to look what the mirror reflects back.	Merriam-Webster Dictionary; Cambridge Dictionary	Literature
"Reflection as an attribute of matter."	Katvan Z, 1978	Philosophy
"Optics seems to know a lot about mirrors."	Eco U, 1986	Literature
"When Marcellus withdrew them [his ships] a bow-shot, the old man [Archimedes] constructed a kind of hexagonal mirror, and at an interval proportionate to the size of the mirror he set similar small mirrors with four edges, moved by links and by a form of hinge, and made it the centre of the sun's beams … Afterwards, when the beams were reflected in the mirror, a fearful kindling of fire was raised in the ships, and at the distance of a bow-shot he turned them into ashes."[b]	Tzetzes J, ~twelfth century AD	Science
"Specchio di Archimede."	Parigi G, 1600	Art

"The most powerful mirror is that of parabolical section, lighting anything combustible with its multitude of fiery darts. A spherical mirror only focuses by a hexagon's angle from the centre, but a parabolic section focuses all beams. Despite all the legends about Archimedes and Cardano's mirrors, one can hardly burn anything ten feet away … It is written how Proclus burned the enemy ships of Constantinople with burning glasses. But I will show you an invention that exceeds these of the ancients and of our age also. For this glass burns at an infinite distance and everything in between. It is unworthy to divulge it to ignorant people, so I will speak in riddles."	della Porta G, 1584	Science
"Heliographs which are considered the very first wireless telegraphs using flashes of sunlight reflected by a mirror were employed by Greeks in their many wars of conquest."	Khan N, 2020	Science
"A piece of glass is a terrible monster of complexity."	Feynman RP, 1985	Science
"There is something deeply disconcerting about mirrors … Despite our constant use of mirrors, our nervous systems remain surprisingly ill-equipped to grasp the mechanics of refraction and reflection."	Martinez-Conde S, 2017	Science
Naive beliefs about optics, like naive physics, lead us to make errors.	Bertamini M, 2003	Psychology
"Un jour deux miroirs s'étant rencontrés s'arrêtèrent pour un brin de causette l'un en face de l'autre. – Tu vois quelque chose? demande le premier. – Non, rien du tout, dit l'autre. – Moi non plus."	Tournier M, 1989	Literature/ tales
"C'est le cas dans une image à l'intérieur de laquelle se trouve reproduite cette même image. Il s'agit littéralement d'une image en partie abîmée … Le cas de l'abîme le plus élémentaire et le plus formel, est celui instauré par deux miroirs placés exactement face à face et reflétant chacun le vide de l'autre porté à une puissance incalculable."	Tournier M, 1986	Literature
"Taking three mirrors, place a pair of them at equal distance from you; set the third midway between those two, but farther back. Then, turning toward them, at your back have placed a light that kindles those three mirrors and returns to you, reflected by them all."	Dante Alighieri, 1321	Literature
"Before the mirror." "Cubic metre of infinity" (is) set inside a reflecting cube. Entering the cube, the visitor can only comprehend with his intellect the infinite reflection without any images in the mirrors that look solely at each other, while he can see with his eyes the reflection of his own self multiplied from one mirror to the other on the walls of the reflecting room."	Falciani C, 2015	Art
"Infinity Mirror Rooms."	Kusama Y, 1963	Art

"By dividing the mirror into two parts and gradually moving the two halves so that they form an angle along the axis of their division, the mirror is multiplied."	Pistoletto M, 1978	Art
"Uomo allo specchio rotto." In the Existential Realism the broken mirror represents the impossibility of men to catch the unitary sense of the existence. The image of men becomes then fragmented and split in many selves.	Ceretti M, 1956	Art
"Breaks the mirror." "The multiplication of the mirror through its shattering… relates to the … relationship between the present and memory … The forms in each mirror remain fixed, while the mirrored images continue to change."	Pistoletto M, 2015	Art
"Jeune fille devant un miroir." The girl reaches out to her different selves. "Mirrors multiply personality … and I mean here multiplication is a very positive self as a claim of growth and development."	Picasso P, 1932	Art
The Venetian mirror is a kind of large mirror around which small mirrors are added as decoration[a].		Science
"Ce cadre énorme, disproportionné … est composé d'une quantité de petits miroirs inclinés dans tous les sens … C'est un miroir dérapant qui chasse vers sa périphérie tout ce qui approche de son foyer … Les miroirs de Venise ne sont jamais droits, ils ne renvoient jamais son image à qui les regarde."	Tournier M, 1975	Literature
"Miroirs dans les logis, miroirs chez les Marchands, Miroirs aux poches des Galands, Miroirs aux ceintures des femmes… Tant de miroirs, ce sont les sottises d'autrui, Miroirs, de nos défauts les peintres légitimes."	de La Fontaine J, 1668	Literature
"First story … deals with the mirror and the shards of glass … A mirror that had the property of reducing everything good and beautiful that was reflected in it into practically nothing… For every speck of the mirror had retained the same power as the whole mirror had possessed."	Andersen HC, 1844	Literature/ tales
"While partial reflection by a single surface is a deep mystery and a difficult problem, partial reflection by two or more surfaces is absolutely mind-boggling."	Feynman RP, 1985	Science
"Thomas Herriot and Johannes Kepler discussed the phenomenon of partial reflection and partial transmission of light in diaphanous bodies. The question was, how could an apparently continuous body transmit and reflect at the same time?"	Meinel C, 1988	Science

In a semiconductor laser "the laser mirrors are formed by the cleaved facets of the semiconductor crystal. Because of the high refractive index (of the semiconductor), the semiconductor-air interface provides a reflectivity… sufficient for laser emission."	Koch SW, 2018	Science
An optical resonator (optical cavity) is a system of mirrors (basically two). It defines the direction of the laser emission radiation.	Hecht E, 1987	Science
"A conventional mirror reflects a light beam according to the well-known reflection law, whereas a phase conjugating mirror reverses the light beam back into itself; thus, generating a conjugated or "time-reversed" light wave … (with) many intriguing applications."	Hampp N, 2003	Science
It can happen to have reflections without mirrors, as in the case of electron microscopy mirror images that sometimes reflect things that are not the object of the investigation[a].		Science
"Le Faux Miroir." A representation of how expectations can be disrupted by bizarre juxtapositions, a challenge to question what we see and what we think we know.	Magritte R, 1929	Art
"Is an object reflected in a mirror perceived differently from an object that is seen directly? … Is the unreality of the looking-glass world reflected in the way that we interact with pictures that do and do not contain reflected objects? … Mirrors produce some perceptual distortions … The difficulty we have in the understanding of mirror reflections … People are undisturbed by paintings with physically impossible depictions of people looking at themselves in mirrors … People discount mirror information even when the mirror provides information not otherwise present in the scene."	Sareen P, 2015	Psychology
The Venus effect is a common phenomenon in which "the mirror itself is used (deliberately or not) to lead us down the wrong path."	Bertamini M, 2003	Psychology
"Rokeby Venus." The size of Venus' face is "at least twice the size it should be."	Velazquez D, 1647 Gregory RL, 1997	Art Psychology
"Marcia painting self-portrait using mirror."	Anonymous, ~1404	Art
Who sees the human face correctly: the photographer, the mirror, or the painter?	Picasso P	Art

"Le miroir." "What makes it really interesting is that we see a sleeping woman next to the mirror; not only she doesn't look at the mirror, she even can't do it—she sleeps! Yet the mirror is here, and actively interacts, if not with the woman then with us, allowing us to see—her backside? or her dreamside?"[c]	Picasso P, 1932	Art
"Selbstbildnis mit Spiegel und Staffelei." Painter, painting, reflection.	Gumpp J, 1646	Art
"Triple self-portrait"	Rockwell N, 1960	Art
"Dalí pintando a Gala por detrás."	Dalì S, 1973	Art
The mirror is able to expand the represented scene both showing what is visible and letting us guess what is not visible[a].		Art
"Arnolfini portrait."	van Eyck J, 1434	Art
"Las Meninas."	Velázquez D, ~1656	Art
"Un bar aux Folies Bergère."	Manet È, 1882	Art

[a]Authors' note

[b]"Greek Mathematical Works, Volume II: Aristarchus to Pappus", translated by Thomas I. Loeb Classical Library. Harvard University Press, 1941

[c]"Picasso and the Women (and a few mirrors in between) (Part II)" retrievable at https://artmirrorsart.wordpress.com/2014/01/05/picasso-and-the-women-and-a-few-mirrors-in-between-part-ii/

Subject Index

M. Milani et al., *Bacterial Degradation of Organic and Inorganic Materials*,
https://doi.org/10.1007/978-3-031-26949-3

T

U

V

Index of Names

M. Milani et al., *Bacterial Degradation of Organic and Inorganic Materials*,
https://doi.org/10.1007/978-3-031-26949-3

Printed by Printforce, the Netherlands